Introduction to Ecology

Introduction to Ecology

Curtis Carson

R CALLISTO REFERENCE

www.callistoreference.com

Callisto Reference,
118-35 Queens Blvd., Suite 400,
Forest Hills, NY 11375, USA

Visit us on the World Wide Web at:
www.callistoreference.com

ISBN: 978-1-64116-604-1 (Hardback)

Cataloging-in-Publication Data

Introduction to ecology / Curtis Carson.
    p. cm.
Includes bibliographical references and index.
ISBN 978-1-64116-604-1
1. Ecology. 2. Biology. 3. Environmental sciences. I. Carson, Curtis.
QH541 .I58 2022
577--dc23

Table of Contents

Preface

The field of biology which focuses on the interactions between the biophysical environment and the organisms which dwell in it is known as ecology. It is closely related to the sciences of genetics, ethology and evolutionary biology. This field of science seeks to understand the effect which biodiversity has on ecological function. There are a number of fields which employ principles from ecology such as agroforestry, conservation biology, agriculture, community health, economics and natural resource management. The actively interacting systems which are made up of organisms, their communities as well as the non-living elements of their surroundings are known as ecosystems. The topics included in this book on ecology are of utmost significance and bound to provide incredible insights to readers. Those in search of information to further their knowledge will be greatly assisted by it. The book will serve as a reference to a broad spectrum of readers.

A detailed account of the significant topics covered in this book is provided below:

Chapter 1- The branch of biology which focuses on the interactions between organisms and their biophysical environment is known as ecology. There are several branches of ecology such as human ecology, population ecology and applied ecology. The topics elaborated in this chapter will help in gaining a better perspective about these branches of ecology.

Chapter 2- The community of living organisms along with the non-living components in their surroundings, which interact as a system is known as an ecosystem. There are several types of ecosystems such as mountain ecosystem, desert ecosystem and aquatic ecosystem. The chapter closely examines the key aspects of these ecosystems along with their components to provide an extensive understanding of the subject.

Chapter 3- The flow of energy within a food chain is known as energy flow. The energy flows within the food web from the producers, also known as autotrophs to the consumers, who are also called heterotrophs. This chapter has been carefully written to provide an easy understanding of the varied types of food chains as well as ecological pyramids.

Chapter 4- The subfield of ecology which studies the dynamics of species populations along with the way in which they interact with the environment is known as population ecology. There are different ways to understand the dynamics between populations such as predator-prey model and Allee effect. The diverse applications of these theories and models related to population ecology have been thoroughly discussed in this chapter.

Chapter 5- The sub discipline of ecology which deals with the interactions between plants and other organisms is known as plant ecology. It also focuses on the distribution and abundance of plants and the effect which environmental factors have upon them. The topics elaborated in this chapter will help in gaining a better perspective about the different interactions which take place in the field of plant ecology.

Chapter 6- The branch of ecology which deals with the interactions between animal populations and their environment is known as animal ecology. Some of the different topics studied within this domain are camouflage, parasitism and mutualisms in animals, as well as predator-prey relationships. This chapter discusses in detail these theories and concepts related to animal ecology.

Chapter 7- The variety and variability of life on the planet Earth is known as biodiversity. It measures variations on several levels such as genetic, species and ecosystem level. The protection of animals, natural areas and plants is called conservation. This chapter discusses in detail the theories and methodologies related to conservation and biodiversity.

I would like to make a special mention of my publisher who considered me worthy of this opportunity and also supported me throughout the process. I would also like to thank the editing team at the back-end who extended their help whenever required.

Curtis Carson

Chapter 1

Ecology: An Introduction

The branch of biology which focuses on the interactions between organisms and their biophysical environment is known as ecology. There are several branches of ecology such as human ecology, population ecology and applied ecology. The topics elaborated in this chapter will help in gaining a better perspective about these branches of ecology.

Ecology is the branch of biology that studies how organisms interact with their environment and other organisms. Every organism experiences complex relationships with other organisms of its species, and organisms of different species. These complex interactions lead to different selective pressures on organisms. The pressures together lead to natural selection, which causes populations of species to evolve. Ecology is the study of these forces, what produces them, and the complex relationships between organisms and each other, and organisms and their non-living environment.

Based on structural components and their relationships, ecology and ecosystem can be explained in two ways: structural concepts and functional concepts.

Structural Concepts

The different types of organisms living in a particular environment are not only independent and mutually reactive but also react with the environment. Though organisms of a species maintain uniformity in their structure and functions through having a common gene pool, they have sufficient plasticity to modify themselves according to changing environment by modifications in somatic characters (ecads) or genetic characters (ecotypes).

Due to their activities, organisms modify the environment to make it more congenial for their growth, development, reproduction and dispersal. The modified environment may become less suitable for the community already living in it. This invites another community that also changes the environment may become less suitable for the community already living in it.

This invites another community that also changes the environment further beyond its most favorable limit. The development of different communities over a period of time at the same site is called succession. The process of succession and change in environment would continue till an equilibrium is established between the changed environment and a community called climax community.

Under similar climatic conditions, different types of communities grow. Some of them have reached their climax stage while others occur in different stages of succession. The complex of many communities growing in a particular area and sharing a common climate is called biome.

Functional Concepts

The biological community consists of a number of organisms and/or populations. Each population occupies a specific volume of the habitat circumscribed by the interaction of various environmental factors and trophic level of the organisms.

It is called ecological niche. The degree of success of a particular population in an area is determined by the parameters of both abiotic factors as well as interaction with other types of populations. The interactions amongst the populations can be positive, negative or neutral.

The flow of energy in the ecosystem is unidirectional or non-cyclic. Radiant energy is trapped by autotrophic plants or primarily producers. From there the energy is transferred to consumers and decomposers. Energy is lost during its transfer from one trophic level to the next. Organisms use the energy in respiration.

A number of inorganic substances are taken by the living beings for their metabolism and body building. They are called biogenetic nutrients. The biogenetic nutrients keep on circulating between the biotic and abiotic components of the ecosystem.

The phenomenon is called biogeochemical cycling. Human beings exploit the ecosphere for their own benefits. As a result, only the economically important plants are allowed to grow in an ecosystem. Species diversity and natural interactions amongst the various components are reduced. When neglected, such an ecosystem deteriorates.

A disturbed or deteriorated ecosystem shows changes due to interactions inside, the assemblage of living being and their abiotic environment, modifying and changing both abiotic and biotic components. The change continues till a stable climax community develops. Where a disturbance continues, the deteriorated ecosystem changes the environment completely.

Ecological Disturbance

Ecological Disturbance is an event or force, of nonbiological or biological origin, that brings about mortality to organisms and changes in their spatial patterning in the ecosystems they inhabit. Disturbance plays a significant role in shaping the structure of individual populations and the character of whole ecosystems.

Minor disturbances include localized wind events, droughts, floods, small wildland fires, and disease outbreaks in plant and animal populations. In contrast, major disturbances include large-scale wind events (such as tropical cyclones), volcanic eruptions, tsunamis, intense forest fires, epidemics, ocean temperature changes stemming from El Niño events or other climate phenomena, and pollution and land-use conversion caused by humans. The notion of ecological disturbance has deep historical roots in ecological thinking; the first conceptual disturbance-related model in modern ecology was ecological succession, an idea emphasizing the progressive changes in ecosystem structure that follow a disturbance.

Characteristics of Disturbance and Recovery

The ecological impact of a disturbance is dependent on its intensity and frequency, on the spatial distribution (or the spatial pattern) and size of the disturbed patches, and on the scale (the spatial extent) of the disturbance. These characteristics are further influenced by the season in which the disturbance occurs, the history of the disturbed site, and the site's topography.

Disturbance Intensity and the Pace of Recovery

The change a terrestrial ecosystem experiences as it recovers from a disturbance depends on the intensity and magnitude of the disturbance. The major mechanisms of recovery in such ecosystems are primary and secondary succession. Primary succession occurs in a landscape that previously was devoid of life. For example, following the retreat of the ice sheets in North America and Eurasia, plants invaded, and a biological recovery was initiated across regions that once had been incapable of sustaining life. In secondary succession, which follows a disturbance in an area with existing communities of organisms, biological remnants (such as buried seeds) survive, and the recovery process begins sooner. The specific identity of these biological "legacies" is dependent on the intensity of the disturbance. For example, the blast from the 1980 eruption of Mount St. Helens devastated some 500 square km (some 200 square miles). Some areas were effectively sterilized, but in other areas organisms survived underground or in patches covered by snow.

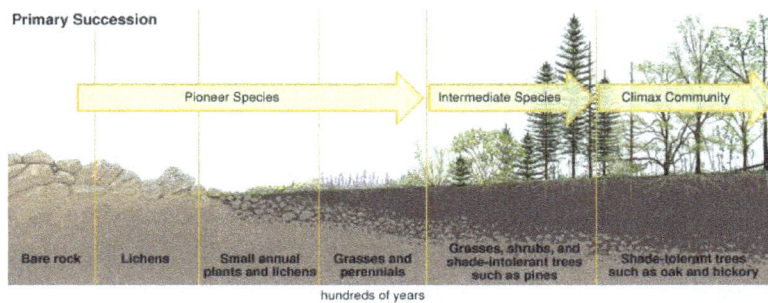

Primary succession

Primary succession begins in barren areas, such as on bare rock exposed by a retreating glacier. The first inhabitants are lichens or plants—those that can survive in such an environment. Over hundreds of years these "pioneer species" convert the rock into soil that can support simple plants such as grasses. These grasses further modify the soil, which is then colonized by other types of plants. Each successive stage modifies the habitat by altering the amount of shade and the composition of the soil. The final stage of succession is a climax community, which is a very stable stage that can endure for hundreds of years.

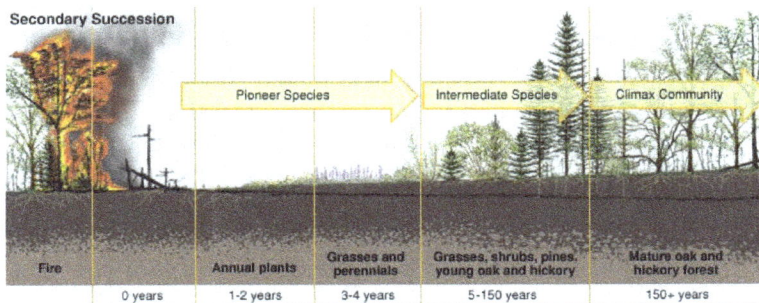

Secondary succession

Secondary succession follows a major disturbance, such as a fire or a flood. The stages of secondary succession are similar to those of primary succession; however, primary succession always begins on a barren surface, whereas secondary succession begins in environments that already possess soil. In addition, through a process called old-field succession, farmland that has been abandoned may undergo secondary succession.

Although the complex mechanisms of succession in marine ecosystems are not well understood, the recovery of these ecosystems is likewise affected by disturbance intensity. For example, beds of giant kelp (Macrocystis pyrifera) that were devastated by the El Niño episodes of 1982–83 and 1997–98 eventually recovered. However, some of these communities needed to be recolonized by propagules, spores in this case (other kinds of propagules include seeds and eggs), coming from other beds hundreds of miles away. Other kelp beds that experienced the effects of lesser El Niño events suffered minimal damage and recovered quickly, because most of the kelp community remained intact.

In both terrestrial and marine ecosystems, the spatial scale of natural disturbances, which is known to span about 10 orders of magnitude, is important. For example, a drought that might devastate protozoans in a temporary pond would be inconsequential to an elephant. A single tree uprooted by a hurricane is a disaster for the resident ants, but it may become a necessary resource for forest frogs as sufficient water collects around the root cavity. Likewise, while a fire may decimate wildlife populations and scorch large areas of land, the cones of the northern jack pine (Pinus banksiana), which are tiny in comparison require the presence of fire to open.

Disturbance Frequency and Recovery

If major disturbances occur too frequently or occur multiple times during an ecosystem's recovery period, they create conditions that can lead to the formation of alternative community states. For instance, Jamaican coral reefs were subjected to an extended period of anthropogenic (human-caused) disturbance during the 20th century, which was characterized primarily by the overharvesting of herbivorous fishes and by pollution. Superimposed on this pattern of persistent degradation were a series of large natural disturbances, including intense hurricanes in 1980 and 1988 and the near-total die-off of Diadema antillarum (a species of sea urchin) that began in 1983. The Jamaican reefs have yet to return to their former coral-dominated state and currently are "fouled" by extensive concentrations of benthic algae in the photic zone. This crowding of the sunlight-infused layer of the ocean creates an alternative state for the ecosystem that can only support small numbers of D. antillarum.

Conversely, disturbance-dependent species suffer when disturbance frequency declines. The unusual sea palm (Postelsia palmaeformis) is a kelp found on marine rocky shores of North America that are exposed to extreme wave scouring. Winter waves produce patches or gaps in the surrounding beds of the California mussel (Mytilus californianus). If these bouts of winter disturbance are frequent enough, the sea palm flourishes. However, at sites characterized by minimal or infrequent wave scouring, it is absent.

The biology of pin cherries (Prunus pensylvanica) illustrates an extension of this theme. In the course of secondary succession in forests of the eastern United States and southern Canada, these small trees grow into gaps and are abundant for periods of about 10 to 25 years; over time, however, as secondary succession progresses, they are competitively eliminated. During the interval of their abundance, their seeds are dropped on the forest floor, where they accumulate and become buried, forming a seed bank. The seeds retain their viability for about 50–150 years. Major disturbances, such as hurricanes or forest fires, cull competitors and thereby permit pin cherries to flourish beyond their typical abundance and replenish the seed bank. Beyond 75 years, however, pin cherry seed viability and seed density diminishes.

Spatial Distribution

Within a given landscape or ecosystem, spatial and biological disturbances can create a mosaic of habitat patches separated by varying distances. The recovery process for species removed by a disturbance is critically dependent on its dispersal capability and the distance between the disturbed site and surviving source populations. For instance, the seeds of many trees are too large to be transported great distances, so their ability to recolonize a disturbed site is measured in metres per generation, rather than kilometres per generation. For some marine invertebrates and algae, however, this distance may be limited to centimetres. For instance, some invertebrate species (e.g., sponges, anemones, snails, and clams) have larvae that crawl short distances. In addition, the spores of some benthic algae are denser than seawater and sink quickly to the bottom. However, propagule transport can span long distances for fugitive or "weedy" species, which are specially adapted to invade and thrive in disturbed environments. In terrestrial environments, adaptations include the development of barbs and hooks (which stick to the fur of mammals), fruits (whose seeds are partially digested by birds and mammals and excreted later), and airfoils (which help the seed glide through the air). In marine systems, spores of green algae and even some floating but fertile plants can traverse great distances.

The fundamental traits of fugitive species—excellent dispersal, high reproductive output, and a brief lifetime—compensate for their reduced competitive prowess. For example, a large disturbance, such as a large wildfire or major wind event, could cut across a forest dominated by beech (Fagus) and maple (Acer), separating what was once a single continuous area into two or more distinct patches. Although weedy species would quickly colonize the disturbed area, the subsequent colonization by larger, hardier tree species would eventually shade out the early arrivals. Several years later, members of the forest's climax community (that is, the final, stable assemblage of plants that is not shaded out by hardier species), which is often composed of mature beeches and maples, would rise in the disturbed area, outcompeting the other trees there.

Size Distribution of Patches

Forest recovery: Tree colonization of the Fantastic Lava Beds in
Lassen Volcanic National Park, northern California.

Within a single event, such as a severe storm or a forest fire, variation in the size of the disturbed patch may be an important factor for recovery. Because patch size has consequences for the regeneration of the local biological community, the effects of numerous small patches are not equivalent to the effects of a larger one of similar total area. More specifically, small disturbed patches can recover rapidly because organisms can easily invade them through short-distance migrations. On the other hand, large patches will persist longer as land or seascape features. They require more time to invade, and their recovery is dependent on propagules produced elsewhere, especially in places where no remnants of the disturbed community survived.

Intermediate Disturbance Hypothesis

Some ecologists claim that these qualitative traits—namely, the persistence of large disturbed patches and the relatively rapid recovery of smaller ones—may be synthesized through the intermediate disturbance hypothesis. This hypothesis states that a disturbance regime (or pattern of disturbances) characterized by low frequency, limited gap size (that is, habitats containing only small areas cleared by disturbances), and low intensity reduces resource availability for many species. Consequently, the variety of species that can coexist locally declines. At the opposite extreme of the disturbance continuum, large-scale and frequent disturbances can restrict community development and the natural evolution of the community. Thus, the hypothesis implies that maximum species richness (i.e., the number of species in a given area) occurs in locations characterized by disturbances whose intensities and frequencies occur at intermediate levels.

Other ecologists contend that the intermediate disturbance hypothesis is problematic. They question the accuracy of the definition of a disturbance and, thus, how one can recognize whether a disturbance has in fact taken place, as well as the most appropriate scale for studying the disturbance. Some ecologists also note that the intermediate disturbance hypothesis overemphasizes species diversity as a measure of recovery from a disturbance. They suggest that other measures, such as the status of various species that suffer or benefit from the effects of the disturbance or the change in the relative abundance of species affected by the disturbance, are more important.

Habitat

Habitats form a vast tapestry of life across the Earth's surface and are as varied as the animals that inhabit them. They can be classified into many genres—woodlands, mountains, ponds, streams, marshlands, coastal wetlands, shores, oceans, etc. Yet, there are general principles that apply to all habitats regardless of their location.

A biome describes areas with similar characteristics. There are five major biomes found in the world: aquatic, desert, forest, grassland, and tundra. From there, we can classify it further into various sub-habitats that make up communities and ecosystems.

Aquatic Habitats

The aquatic biome includes the seas and oceans, lakes and rivers, wetlands and marshes, and lagoons and swamps of the world. All of these habitats are home to a diverse assortment of wildlife. It includes virtually every group of animals, from amphibians, reptiles, and invertebrates to mammals and birds.

The intertidal zone, for instance, is a fascinating place that is wet during high tide and dries up as the tide goes out. The organisms that live in these areas must withstand pounding waves and live in both water and air. It is where you will find mussels and snails along with kelp and algae.

Desert Habitats

Deserts and scrublands are landscapes that have scarce precipitation. They're known to be the driest areas on Earth and that makes living there extremely difficult. Deserts are rather diverse habitats. Some are sun-baked lands that experience high daytime temperatures. Others are cool and go through chilly winter seasons. Scrublands are semi-arid habitats that are dominated by scrub vegetation such as grasses, shrubs, and herbs.

It is possible for human activity to push a drier area of land into the desert biome category. This is known as desertification and is often the result of deforestation and poor agricultural management.

Forest Habitats

Forests and woodlands are habitats dominated by trees. Forests extend over about one-third of the world's land surface and can be found in many regions around the globe.

There are different types of forests: temperate, tropical, cloud, coniferous, and boreal. Each has a different assortment of climate characteristics, species compositions, and wildlife communities.

The Amazon rain forest, for example, is a diverse ecosystem, home to a tenth of the world's animal species. At almost three million square miles, it makes up a large majority of Earth's forest biome.

Grassland Habitats

Grasslands are habitats that are dominated by grasses and have few large trees or shrubs. There are two types of grasslands: tropical grasslands (also known as savannas) and temperate grasslands.

The wild grass biome dots the globe. They include the African Savanna as well as the plains of the Midwest in the United States. The animals that live there are distinct to the type of grassland, but often you'll find a number of hooved animals and a few predators to chase them. Grasslands experience dry and rainy seasons. Due to these extremes, they are susceptible to seasonal fires and these can quickly spread across the land.

Grasslands are habitats that are dominated by grasses and have few large trees or shrubs.

Tundra Habitats

Tundra is a cold habitat. It is characterized by low temperatures, short vegetation, long winters, brief growing seasons, and limited drainage. It is an extreme climate but remains the home to a variety of animals. The Arctic National Wildlife Refuge in Alaska, for instance, boasts 45 species ranging from whales and bears to hearty rodents.

Arctic tundra is located near the North Pole and extends southward to the point where coniferous forests grow. Alpine tundra is located on mountains around the world at elevations that are above the tree line.

Systems Ecology

Systems ecology is then the study of ecosystems that uses mathematical modeling, computation and, as the name implies, is based on systems theory. As with other areas of systems science, the use of systems theory as an approach involves the adoption of a holistic paradigm based on synthetic reasoning, meaning that systems ecology seeks a holistic view of the interactions and transactions within and between biological and geological systems on various scales. With this alternative approach, it does not restrict itself simply to the study of natural biophysical processes, but systems ecology now gives equal attention to the human dimension. Whereas standard ecology sees human industrial and economic activity as largely outside of its domain, systems ecologists recognize that the function of any ecosystem can be influenced by human economics in fundamental ways and that human industrial economic activity is a fundamental part of ecosystems around the world today. It has therefore taken an additional transdisciplinary step by including economics in the consideration of coupled socio-ecological systems. As such systems ecology takes an expansive domain of interest crossing almost all areas, from physics to biology, to economics and social studies to truly try and understand the workings of earth's systems in all their multi-dimensional complexity.

Analysis

A central part of systems ecology is the holistic paradigm derived from systems theory, which is in contrast to our more traditional approach taken within the natural sciences called analysis. Traditionally within modern science when looking at the macroscopic features of a given system, scientists have tried to find the origin of these phenomena by looking at the structure and properties of their component parts, by breaking the system down and then describing it as some linear combination of the parts, this process of reasoning is called analysis. Most of the success of modern science has relied on an analytical reductionist approach in which systems are taken apart to examine the individual components and how they interact together.

Historical examples are the isolation and characterization of the elements of the periodic table and the discoveries of the particles that make up atoms. In the biological sciences, reductionism has also been very successful: examples range from the purification of proteins, DNA and RNA and the study of their structures and activities, to the sequencing and analysis of whole genomes.

Complexity

In the past, the usual way to study complex phenomena was based on simplifying them through analytical reductionism, describing them as simple systems analogous to machines or by aggregating and averaging through statistical analysis, describing them as unorganized complexity. But complex systems, such as ecosystems, exist at a threshold between order and chaos because they are too complex to be treated as machines and too organized to be assumed random and averaged, they are best understood in terms of patterns and processes that emerge as we put the parts together. Simple systems may be governed by a single global rule that can be described in a beautifully compact equation, but complex systems are not governed by a single rule, they are what emerges out of the distributed interaction of many elements.

Synthesis

An ecology is what emerges out of the interaction of many different biotic and abiotic elements on different levels. Where as with analysis we are breaking physical systems down to their most basic constituent elements, with systems ecology we are interested in what happens when we put things together, the processes that emerge on different levels as we build them up. Instead of talking about the properties of parts, we are talking about the connections between them. The fact that the properties of the individual units cannot always explain the whole has been known from the earliest times of science. In this context, it is often said that the whole is more than the sum of the parts, meaning that the global behavior exhibited by a given system will display different features from those associated with its individual components.

Irreducibility

A more appropriate statement would be that "the whole is something else than the sum of its parts" since in most cases completely different properties arise from the interactions among components. As an example, the properties of water that make a molecule so unique for life cannot be explained in terms of the separate properties of hydrogen and oxygen, even though we can understand them in detail from quantum mechanical principles. Some properties such as memory in the brain cannot be reduced to the understanding of single neurons. Life itself is a good example: nucleic acids, proteins, or lipids are not "alive" by themselves. It is the cooperation among different sets that actually creates a self-sustained, evolvable pattern called life.

Systems Biology

An example of this new approach to science is systems biology, which recognizes that through the analytical approach we have gained a very thorough understanding of the component parts of biological systems. There is a large and rapidly growing body of information about the building blocks of cells, proteins, RNA, DNA, lipids etc. but how these molecules form organelles, and how cells form tissues and organisms, is far from understood, or equally how in developmental biology, the genome creates the organism? Self-organization plays a central role in all these processes and the answers are still largely a mystery. This is obviously not just an issue in biology on the micro level but also on the macro level in understanding ecosystems and just as importantly coupled socio-ecological systems.

Systems Theory

Systems ecology studies ecosystems through abstract mathematical models and computation. An ecosystem model is an abstract, usually mathematical, representation of an ecological system ranging in scale from an individual population, to an ecological community, or even the entire biosphere, which is studied to gain understanding of the real system. Systems theory is a formal modeling language that is based on the model of a system, a system is a highly abstract model, in its essence it is simply a set of parts and relations between those part through which they are interdependent in effecting some joint outcome.

This model is very effective in providing a generic language for talking about all kind of entities from a single cell to the entire Earth system. Systems ecology often deals with ecosystems on a higher level of abstraction than standard ecology, in order to be able to remove the details and derive formal models. These formal models of systems theory go hand in hand with computational methods that enable the interpretation of large amounts of data. This approach of using abstract models and computation allows us to approach understanding very complex ecosystems, such as the whole ecology of Earth in a formal fashion. Systems ecology is one of the few theoretical tools that can simultaneously examine a system from the level of individuals all the way up to the level of ecosystem dynamics. It is an especially valuable approach for investigating systems so large and complex that experiments are impossible, and even observations of the entire system are impractical.

Energetics

A second fundamental set of ideas within systems ecology is that of energetics, interpreting ecosystems in terms of the flow of energy through networks. Systems ecology studies the flow of energy and materials through networks of biotic and abiotic elements within ecosystems. It seeks to understand the processes which govern the stock and flow of material and energy and how they are processed through the system. Any ecosystem is characterized by flows: flows of nutrients and energy, flows of materials, and flows of information. It is such flows that provide the interconnections between parts, and transform the community from a random collection of species into an integrated whole, an ecosystem in which biotic and abiotic parts are interdependent. The analysis of how ecosystems function is determined by how those processes and components cycle, retain, process and exchange energy and nutrients. Systems ecology typically involves the application of computer models that track the flow of energy and materials and predict the responses of systems to perturbations.

Ecosystems and biological systems in general challenge us because they are constantly consuming energy, and are therefore far from thermal equilibrium. Thus classical thermodynamics, which has been so successful in developing an understanding of physical and chemical properties such as temperature and pressure, does not apply to these systems. Instead of self-assembling into the lowest energy state, such as a crystal, these energy-dissipating components self-organize into highly dynamic structures, through which there is a constant flux of energy and material, and this is in many ways the defining feature of life. Within chemistry, this is called a dissipative system and the theory of dissipative systems goes a long way to helping us understand how biological systems self-organize and evolve over time into more complex organizations.

Emergence and Hierarchy

The idea of hierarchy and integrative levels of organization is another major organizing theme within systems ecology. Integrative levels is an extension of the idea of emergence that addresses the biological organization of life that self-organizes into layers of emergent whole systems that function according to non-reducible properties. This means that higher order patterns of a whole functional system, such as an ecosystem, cannot be predicted or understood by a simple summation of the parts. These hierarchical structures have a nested pattern where smaller subunits are nested within larger subsystems and so on.

Emergence gives ecosystems a distinctive omnipresent hierarchical structure and this scale hierarchy is a primary organizational principle, from biological cell to individual to community to ecosystem to biosphere. The study of ecosystems can cover 10 orders of magnitude, from microbes on the surface layers of rocks to the surface of the planet. In this hierarchy there are both processes and patterns that are universal, having a scale-invariant, fractal property as they recur on all levels, but also unique processes emerge on the different levels. This idea of synergies and self-organization leading to emergence and the formation of new levels in a hierarchical fashion is a central model for understanding the complex multi-dimensional characteristic of ecosystems.

Feedback Loops and Homeostasis

Another major modeling approach adopted from cybernetics and systems theory is that of feedback loops, which are central to understanding the dynamics of macro scale complex systems as they evolve over time and also to understanding processes of regulation and control within ecosystems and economy. On the micro-level feedback is well understood in the process of homeostasis, which means maintaining things at a steady state, negative feedback homeostasis is used in biochemical processes to regulate cells, individual organs and organisms. But macro processes of change, such as ecological succession are also regulated by feedback loops, as we go up from the organism to the community and the whole biosphere there is no homeostatic centralized control system, but now instead distributed feedback loops that work to stabilize the macro system into an oscillatory flow, bound within some upper and low limits. This process is called homeorhesis a term derived from the word "same" and "flow" as it refers to a stabilized flow.

Feedback loops tell us a lot about the dynamics to ecosystems and biological systems in general, for example negative feedback regulates the human body as it grows. Starting with what is called a R stage of growth where most of the resources are used for development and little for maintenance, a period of high growth rate and positive feedback. Before at some stage reaching a mature state where negative feedback starts to limit the growth as the system enters what is called a k stage of growth, investing resources in other activities with negative feedback setting in as the system becomes more mature. Feedback loops are an example of nonlinear models that can be used to understand complex behavior within both ecosystems economies and the interaction be-tween them.

Human Ecology

Human ecology is about relationships between people and their environment. In human ecology the environment is perceived as an ecosystem. An ecosystem is everything in a specified area - the air, soil, water, living organisms and physical structures, including everything built by humans. The living parts of an ecosystem - microorganisms, plants and animals (including humans) - are its biological community.

Ecosystems can be any size. A small pond in a forest is an ecosystem, and the entire forest is an ecosystem. A single farm is an ecosystem, and a rural landscape is an ecosystem. Villages, towns and large cities are ecosystems. A region of thousands of square kilometres is an ecosystem, and the planet Earth is an ecosystem.

Although humans are part of the ecosystem, it is useful to think of human - environment in-teraction as interaction between the human social system and the rest of the ecosystem. The social system is everything about people, their population and the psychology and social orga-nization that shape their behaviour. The social system is a central concept in human ecology because human activities that impact on ecosystems are strongly influenced by the society in which people live. Values and knowledge - which together form our worldview as individuals and as a society - shape the way that we process and interpret information and translate it into action. Technology defines our repertoire of possible actions. Social organization, and the social institutions that specify socially acceptable behaviour, shape the possibilities into what we actu-ally do. Like ecosystems, social systems can be on any scale - from a family to the entire human population of the planet.

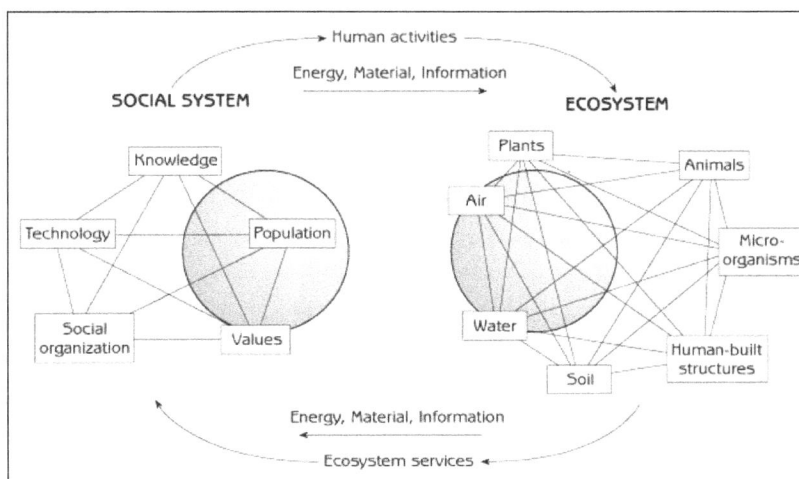

Interaction of the human social system with the ecosystem

The ecosystem provides services to the social system by moving materials, energy and information to the social system to meet people's needs. These ecosystem services include water, fuel, food, materials for clothing, construction materials and recreation. Movements of materials are obvious; energy and information are less so. Every material object contains energy, most conspicuous in foods and fuels, and every object contains information in the way it is structured or organized. Information can move from ecosystems to social systems independent of materials. A hunter's discovery of his prey, a farmer's observation of his field, a city dweller's assessment of traffic when crossing the street, and a refreshing walk in the woods are all transfers of information from ecosystem to social system.

Material, energy and information move from social system to ecosystem as a consequence of human activities that impact the ecosystem:

1. People affect ecosystems when they use resources such as water, fish, timber and livestock grazing land.

2. After using materials from ecosystems, people return the materials to ecosystems as waste.

3. People intentionally modify or reorganize existing ecosystems, or create new ones, to better serve their needs.

With machines or human labour, people use energy to modify or create ecosystems by moving materials within them or between them. They transfer information from social system to ecosystem whenever they modify, reorganize, or create an ecosystem. The crop that a farmer plants, the spacing of plants in the field, alteration of the field's biological community by weeding, and modification of soil chemistry with fertilizer applications are not only material transfers but also information transfers as the farmer restructures the organization of his farm ecosystem.

Population Ecology

Population ecology is the study of these and other questions about what factors affect population and how and why a population changes over time. Population ecology has its deepest historic roots, and its richest development, in the study of population growth, regulation, and dynamics, or demography. Human population growth serves as an important model for population ecologists, and is one of the most important environmental issues of the twenty-first century.

Properties of Population

1. Dimensions - population density, abundance or size, based on the number of individuals (per unit area, or in the entire population);

2. Composition - the proportions of individuals with different states (gender, age, developmental stage, or size);

3. Dynamics - changes in time of the dimensions and composition of a population; this includes the rate of change of the dimensions (population growth rate), and changes in the distributions of individuals' states.

The properties of populations change all the time and the rate of change can vary from very low to high. Even in populations that do not change in density, the individuals change and are replaced. They are born, grow and reproduce (or not), and die. These changes in the state of individuals can be summarized as rates of change, such as birth rate and mortality rate, or the proportion of individuals growing to a particular size in an interval of time, etc. These demographic processes are direct causes for the changes (or lack of it) in the population's dimensions, composition and dynamics.

Demographic processes are abstractions, derived from actual changes in the state of the individuals. They are calculated as the proportion of individuals in a particular stage who manage to get to the next stage. Each individual experiences a different rate of change, in accordance with its general condition, its past and its genotype. Vital rates are demographic processes averaged over groups of individuals (according to age, size, or stage). Vital rates include birth rate, survivorship, growth, and fertility or fecundity. The states of individuals and the rates that form the connections between them, determine the life-cycle of the population.

Population and Environment

Two different populations of the same species differ first of all in their vital rates, since they express how individuals respond to environmental variation. For example, comparing a population of a particular animal species in a rich and a poor habitat regarding food availability, we will usually see that population growth rate in the rich place is higher than in the poorer place. This is because individuals survive better, grow faster, attain reproductive maturity faster, and produce more offspring when there is more food.

Environmental factors that affect demographic processes are very diverse and can be divided in many ways. Categories are not exclusive: some factors may appear in more than one category. The following division is based on the properties of the factors and their effects on the organisms.

1. Environmental conditions - states of the environment that affect organisms, but are not consumed by them: pH, temperature, air humidity (including their variation).

2. Resources - physical commodities (material or energy in any form, including other organisms), which are supplied by the environment and consumed by the organism; resources are at least as diverse as the organisms that require them.

3. Structures - availability of sites where individuals can perform their essential functions (shelter, nesting, foraging, mating and recruitment sites); sites can be viewed as a structural sort of resource.

4. Interactions with other organisms - all relationships with other organisms except consuming another as resource; interactions include (the passive side of) predation, herb ivory and parasitism, and competition for resources and structures; besides negative interactions, they can be positive as well (facilitation, mutualism).

5. Internal regulation - effects of the population itself (or some of its members) on the way environmental factors affect the organisms. (Strictly speaking this is not an environmental factor, although for each member of a population the other members are part of its

environment.) Processes affected this way are called density-dependent. Density-dependence can be positive or negative, depending on whether density of the population reduces or enlarges the effect of the environmental factor. Intraspecific competition is a negatively density-dependent process, and operates on resource acquisition.

Plant Ecology

Plant ecology examines the relationships between plants and their physical and biotic environment. Plants are mostly sessile and photosynthetic organisms, and must attain their light, water, and nutrient resources directly from the immediate environment. Plant size and position in the community affect the capture and utilization of these resources and hence plants have evolved specific adaptations to enhance these capabilities. Understory plants have evolved mechanisms that allow them to tolerate low light conditions, while plants in the open have different mechanisms to cope with excess light. The absorption by roots and movement of water in the plant are determined by gradients in potential energy between the soil and atmosphere, as well as within the plant, as expressed by the concept of water potential. Nutrients are available through biological and chemical processes in the soil. Mycorrhizae are critical in absorption of phosphorus and are also capable of interconnecting plants through their hyphae, thus facilitating belowground transfers of nutrients and water. Plants possess various adaptive functions, such as different photosynthetic pathways, that provide greater fitness in certain environments. In addition, there are correlations among plant traits, such as a positive relationship between photosynthetic rate and leaf nitrogen, or between leaf mass per area and photosynthesis, which suggest that there are ecological rules governing functional traits that cross species lines. Resource competition occurs when one or more resources are in limited supply and plants have various adaptations that maximize competitive success, including. Allelopathy, when one plant releases an organic material into the environment to the detriment of a second plant. Plants also greatly influence the belowground environment (the rhizosphere) by altering the composition of the microbial community of bacteria and fungi. Interactions between above and belowground processes affect competitive outcomes and can alter community dynamics, including the process of successional change. Primary succession occurs on new substrate and secondary succession occurs where vegetation previously existed. Secondary successions are initiated by disturbances such as fire, wind damage, flooding, grazing, and disease. Disturbance frequency and intensity greatly determine the development of the plant community and current and future climate change may result in new communities not present under present conditions, nor that resemble any from the recent past, making predictions of such impacts difficult.

Animal Ecology

Animal ecology is the interrelationships between animals and their biotic and physical environments, emphasizing animal population and community dynamics. Its study consist of areas such as animal species and their taxonomic relations (systematics), animal distribution and abundance within their range (zoogeography), interaction between species, individuals and their environment (ecology) and whole animals as functional units (physiology).

Ecology and Habitats

Animals evolved in the seas but moved into fresh water and onto land in the Ordovician Period, after plants became available as a food source. A simple history of animal ecology centres on the theme of eating some organisms for food while providing food for others. The realities of how animals have done so are richly varied and complex. The ecology of animals and other organisms is reflected in their phylogenetic radiations (i.e., the diversification of lineages). Ecologies are as numerous as species, but, just as species can be grouped into higher taxa, so too can a classification be made of the ways by which animals find adequate food to reproduce and the ways they remain alive while doing so.

Competition and Animal Diversity

The majority of animal phyla are, and have always been, confined to the sea, a comparatively benign environment. Marine animals need not osmoregulation, thermo regulate, or provide against desiccation. The energy procured can thus be used mostly for growth, reproduction, and defence. Even reproduction can be simple: shunting millions of eggs and sperm into the water and letting them fend for themselves. Developing embryos do not need the protection of a womb because the ocean provides a suitable environment.

Despite the simplicity an animal's life can attain within the ocean, most oceanic animals have not remained simple. Competition and predation, two major components of any habitat, have complicated the lives of animals, leading to ever more novel ways of surviving. No matter how inimical to life, the physical components of environments are relatively predictable elements to which adaptation is often comparatively easy, if costly. Competition and predation, in contrast, relentlessly challenge all forms of life no matter how perfect they become for an instant in time. Adaptations often become obsolete as soon as they are successful, because successful life forms become a prime source of food for others.

Given the simple thesis that competition drives much adaptation, the ecological diversity of animals can be sketched readily. Form, function, and phylogenetic history reflect the roles that animals assume in the evolutionary drama. Throughout a billion-year history, the animal actors have changed many times, but they perform variations on the same theme and the backdrops look much the same. For example, shortly after plants became well established, forests of giant lycopods (club mosses) and tree ferns provided food and shelter for numerous arthropods, including winged insects, on which four-legged amphibious vertebrates fed. Larger amphibians and reptiles later turned to smaller ones for food. Some of the arthropods and other terrestrial animals in turn were parasitic on the vertebrates. Later, different groups of plants, insects, and vertebrates enacted the same scene. First gymnosperms and then angiosperms became the dominant components of forests. Amphibians yielded dominance to mammal-like reptiles (some of which became herbivorous), which gave in to dinosaurs; the latter were replaced by mammals and, most recently, by humans. In aquatic habitats the same drama has unfolded, with ever-changing actors. Reefs, for example, have entirely disappeared several times, with each subsequent avatar built mostly from different kinds of organisms. A historical perspective illustrates the underlying direction provided by competition and predation.

Evolution of Ecological Roles

Animals arose from protozoans and initially were simply larger, more complex, and successful competitors for the same sources of food. The early animals (parazoans, coelenterates, flatworms, and extinct groups) exhibited the same basic strategies of obtaining food as did the protozoans. Because of their larger size, however, they had an advantage over protozoans: they could prey on them and oust them from their attachment sites on the ocean floor. The early basic strategies of animal life reflected two different means of competing for food, that fixed by photosynthetic and chemosynthetic organisms and that provided by the wastes and decaying tissues of life forms. Almost all the free energy fixed is used by one organism or another, so that what one animal wins is lost to the rest. Animals do whatever they can to acquire all the energy they can use, and in this basic sense each is competing with all the others. Ultimately, predation is a mode of competition that simply involves eating the potential competitor rather than finding another way to share the same resource.

Three early ecological roles of animals were as filter feeders, predators, and scavengers. The filtering of comparatively tiny organisms and organic detritus is a form of predation that was easily acquired when an animal became immense relative to potential food. Sponges were the earliest filter-feeding animals and still dominate certain marine habitats.

Predation on relatively large organisms relies on capture and subsequent subjugation of the prey until it can be ingested. Predation grades into filter feeding when the prey is very small in relation to the predator and into parasitism when very large. Among the early animals, coelenterates were the initial predators. Either attached to the bottom or floating near the surface, they paralyzed potential prey with their stinging and muscular tentacles and pushed it into their guts to digest it at leisure. Placozoans and flatworms preyed somewhat less effectively; they crept over a sessile or slow-moving potential prey, formed a pocket around it, and then ejected digestive enzymes to break it into smaller pieces that could be ingested.

Scavengers feed on the remains of dead organisms. A layer of energy-rich organic matter continuously settles on the ocean bottom, where it is recycled by diverse organisms. As animals evolved, they became essential as garbage eliminators because their remains (and those of plants and some fungi) are only slowly decomposed by microorganisms. Without animal scavengers, ocean bottoms and land surfaces would be cluttered with the refuse of dead organisms. Among the early animals, flatworms were the primary scavengers on the ocean bottoms.

Although the early radiation of animals admirably filled the major ecological roles, they had structural, physiological, and behavioral limitations that left some options open. For example, there were no potential predators of the surface-creeping flatworms or placozoans or of the cnidarians, with their stinging cells. There were no burrowers that could penetrate the layer of detritus which undoubtedly accumulated on the ocean floor. With the acquisition of a coelom or pseudo coel, animals could burrow into the detritus layer, consuming it as they went, as earthworms do on land.

Well-developed organ systems permitted an increase in body size, which gave rise to successive levels of predators. Quite early in the rapid diversification of animal life, protective hard shells appeared, a defence against predators but later also a means of enabling animals to expand outward from the seas. The intertidal areas, with partial exposure to the atmosphere, became a liveable habitat. Jaws were an important innovation to predators. They are particularly central to the overwhelming success of arthropods and vertebrates, especially on land, where most plants and animals possess a tough drought- and injury-resistant covering. Most mollusks have a file like radula that is well suited for breaking down tough plant or animal tissue into ingestible pieces and even adequate for drilling through the thick shells of their own group.

Large size, made possible by rigid skeletal support, particularly for reef-forming animals, provided shelter and thus more variations on the common themes. Corals and some other animals shelter algae, particularly in the nutrient-poor tropical seas, and obtain their food directly from their symbioses. This was probably more common in the Ediacaran Period (the last interval of Precambrian time, from about 635 million to 541 million years ago on certain geologic time scales), when thin animals could bask in the water without predators. Most of the deep sea is sparsely populated, its animals living on what settles down from above, but volcanic and other deep-sea vents emit gases that can be oxidized to provide energy. Some animals have symbiotic bacteria that do this, and they reach high densities there. Photosynthetic reef builders create forests in the seas that are analogous to those on land.

Large body size also favours the rise of parasitism, the consumption of living tissue that typically does not kill the host organism outright. Too heavy a load of parasites weakens an animal and makes it more susceptible to predation or other forms of death. Parasites have evolved in many phyla, the most important being platyhelminths, nematodes, and arthropods. Several taxa of high level are entirely parasitic. A disadvantage of parasitism, particularly on land, is dispersal to another host. Intermediate hosts are sometimes used if direct passage cannot be made. Enormous reproductive output is the rule (other organ systems can be minimal because the habitat is so congenial). The extraordinary number of species of winged insects attests to the success of the parasitic way of life. Insects can actually feed in the dispersal stage and thus survive longer while seeking the appropriate host.

Applied Ecology

Applied ecology is a scientific field that studies how concepts, theories, models, or methods of fundamental ecology can be applied to solve environmental problems. It strives to find practical solutions to these problems by comparing plausible options and determining, in the widest sense, the best management options.

One particular feature of applied ecology is that it uses an ecological approach to help solve questions concerned with specific parts of the environment, i.e., it considers a whole system and aims to account for all its inputs, outputs, and connections. accounting for everything is no more possible in applied ecology than it is in fundamental ecology, but the ecosystem approach of applied ecology is both one of its characteristics and one of its strengths.

Indeed, one could view the overall objective of applied ecology as to maintain the focal system while altering either some of the elements we take from the system (i.e., ecosystem services or exploitable resources) or some of those we add to the system (i.e., exploitation regimes or conservation measures) through an educated management strategy. Since those two types of elements are not mutually independent, long-term management strategies are best aimed at optimizing rather than maximizing exploited items. This is more efficiently achieved through an adequate understanding of theoretical ecology, which generally considers all parts of the system rather than a limited set of its components.

Aspects of applied ecology can be separated into two broad study categories: the outputs and the inputs. The first contains all fields dealing with the use and management of the environment for its eco-system services and exploitable resources. These can be very diverse and include energy (fossil fuel or renew-able energies), water, or soil. They can also be biological resources—for their exploitation—from fish to forests, to pastures and farmland. They might also, on the contrary, be species we wish to control: agricultural pests and weeds, alien invasive species, pollutants, parasites, and diseases. Finally, they can be species and spaces we wish to protect or to restore.

The fields devoted to studying the outputs of applied ecology include agro-ecosystem management, rangeland management, wildlife management (including game), landscape use (including development planning of rural, woodland, urban, and per-urban regions), disturbance management (including fi res and floods), environmental engineering, environmental design, aquatic resources management (including fisheries), forest management, and so on. This category also includes the use of eco-logical knowledge to control unwanted species: biological invasions, management of pests and weeds (including biological control), and epidemiology.

The inputs to an applied ecology problem consist of any management strategies or human influences on the target ecosystem or its biodiversity.

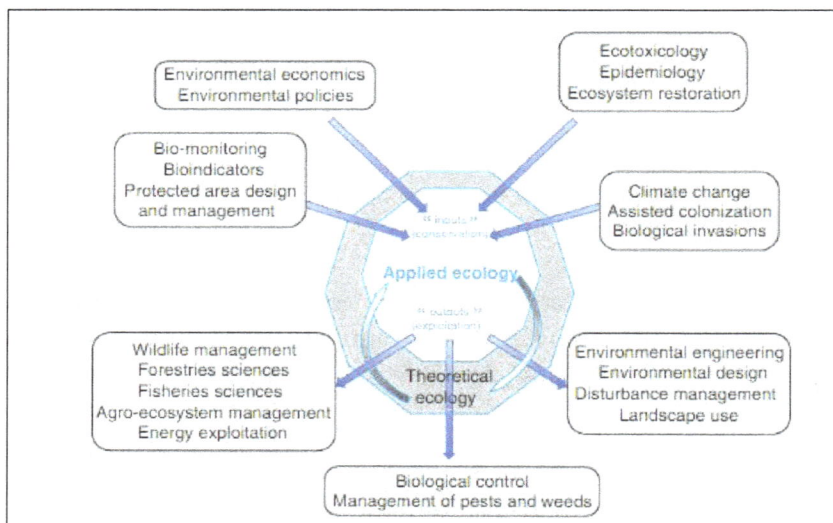

The relationships between the different fields of applied and theoretical ecology.

These include conservation biology, ecosystem restoration, protected area design and management, global change, ecotoxicology and environmental pollution, bio-monitoring and bio-indicators of environmental quality and biodiversity, environmental policies, and economics. these outputs and inputs are intimately connected. For example, the management of alien invasive species is relevant to both natural resource management (e.g., agriculture) and the protection of biodiversity (conservation/restoration biology).

In addition to using fundamental ecology to help solve practical environmental problems, applied ecology also aspires to facilitate resolutions by nonecologists, through a privileged dialogue with specialists of agriculture, engineering, education, law, policy, public health, rural and urban planning, natural resources management, and other disciplines for which the environment is a central axiom. Indeed, some of these disciplines are so influential on environmental management that they are viewed as inextricably interlinked. For example, conservation biology should really be named "conservation sciences" because it encompasses fields that are not very biological, such as environmental law, economics, administration and policy, philosophy and ethics, resources management, psychology, sociology, biotechnologies, and more generally, applied mathematics, physics, and chemistry.

And obviously, as we are dealing with environment and ecosystems, everything is connected, all questions are interrelated, all disciplines are linked, and all answers are interwoven. Understanding a process through one field of applied ecology will allow advancing knowledge in other fields.

Links between Applied and Theoretical Ecology

The links between theoretical and applied ecology range from spurious to robust and have been used with varying success in different fields. Fields that have benefited from theory include fisheries and forestry management and veterinary sciences and epidemiology (both human and nonhuman). However, some other fields of applied ecology have not (or not yet) benefited fully from ecological theories, concepts, and principles, and that aspect is the focus of this entry.

Three such examples are presented here. The first is an illustration of a major field of ecology that has a strong applied branch but has not, until very recently, fully taken advantage of theoretical ecology: invasion biology. The second describes a theoretical process important to many aspects of applied ecology (including epidemiology, fisheries, biological control, conservation biology, and biological invasions) but which has nonetheless been underused so far in applied ecology: the Allee effect. The third depicts an emerging field that is now, by obligation, mostly theoretical but which has an applied future and for which it is hoped that ecological applications will emerge rapidly: climate change.

Paleoecology

Ecology of prehistoric times, extending from about 10,000 to about 3.5×10^9 years ago. Although the principles of paleoecology are the same as those underlying modern ecology, the two fields actually differ greatly. Paleoecology is a historical science that must rely on empirical data from fossils and their enclosing sedimentary rocks to make inferences about past conditions. Experimental approaches and direct measurement of environmental parameters, which are critical components of modern ecology, are generally impossible in paleoecology. Furthermore, distortion and loss of information during fossilization means that fossil assemblages and distributions are rarely congruent with living communities. Hence, the resolution of ancient ecosystems must remain relatively imprecise. The lack of precision is compensated for by the fact that paleoecology deals with processes occurring over vast spans of time that are unavailable to modern ecology. Long-term changes in communities (replacement) may be discerned and related to patterns of environmental change. More significantly, overall patterns of ecological change in the global biosphere may be documented; evolutionary paleoecology focuses on recognition and interpretation of long-term ecological trends that have been critical in shaping evolution.

Among the goals of paleoecology are the reconstruction of ancient environments (primarily depositional environments), the inference of modes of life for ancient organisms from fossils, the recognition of recurring groupings of ancient organisms that define relicts of communities (paleo communities), the reconstruction of the interactions of organisms with their environments and with each other, and the documentation of large-scale and long-term patterns of stasis or change in ecosystems.

Paleoenvironmental Interpretations

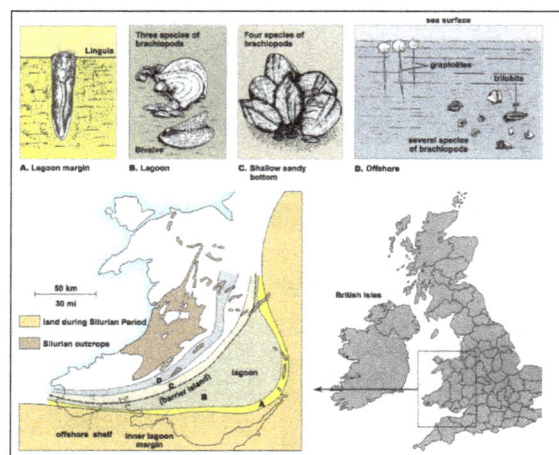

Zonation of Silurian (425 million year old) marine communities in the Welsh Basin.

To reconstruct ancient marine environments, many different parameters must be inferred, such as temperature, water salinity, oxygen levels, nutrient concentrations, and water movements and depth. In this regard, paleoecology interfaces directly with the fields of sedimentology and stratigraphy, including study of modern depositional environments.

Marine animal communities were typically arranged in belts parallel to shoreline and related to water depth. A classic study using fossil communities in basin analysis mapped out the distribution of different communities of marine fossils, primarily brachiopods, to show the contouring of belts of ancient environments from the shoreline in the southeast to deep basinal environments.

Taxonomic Uniformitarianism

One of the most useful, but also potentially misused, aspects of paleoecological application is known as taxonomic uniformitarianism. This concept relies on studies of modern organisms to determine limiting environmental factors, such as salinity tolerance, temperature preference, or depth ranges. Fossils of the same or closely related species are then inferred to have had similar environmental preferences, and their occurrence is judged to indicate that particular strata were deposited under a comparable range of environmental conditions. Such an approach is valid for very closely related organisms in relatively recent geologic time. Species and even genera may have relatively uniform environmental ranges through time, but the same cannot always be said of higher taxa such as families. At the level of order or class, only the broadest uniformitarian generalizations apply. For example, it is probably valid to consider fossil nautiloids or echinoderms as indicators of normal marine salinities, as all living representatives of these taxa have very limited abilities for osmotic regulation and therefore are restricted to near-normal salinity. Similarly, the restriction of photosynthetic organisms (such as algae) to the euphotic zone may be useful in determining relative depth. However, the precision and reliability of taxonomic uniformitarianism breaks down in increasingly ancient samples.

Morphologic Features

Certain features of the morphology of fossils may be useful in making environmental inferences without reliance on evolutionary relationships. For example, the presence of entire margins and drip tips on leaves of plants is indirectly related to humid, warm climates, and so the proportions of leaf floras with entire margins and drip tips have been used as an index of paleolatitudinal zonation. Growth forms of colonial organisms relate to environmental factors such as turbulence and sedimentation rates. Flexible or articulated skeletons or flat encrusting form in colonial marine animals are associated with highly turbulent shallow-water environments where streamlining becomes important. Delicately branched, inflexible colonies typify quiet areas and, commonly, areas of high turbidity where a branching skeleton may shed sediment more readily than a flat or globose form. Such ecologically related morphology may transcend taxonomic boundaries.

Skeletal Mineralogy and Geochemistry

The microstructure and geochemistry of organism skeletons may provide clues about ancient environments. For example, the presence of growth banding in skeletons provides evidence for seasonal variability in climates. The skeletons of fossil organisms, if they are well preserved, also encode valuable environmental information in the form of trace elements and isotopic signatures. For example, the calcium carbonate skeletons of marine invertebrates incorporate trace elements whose proportion is related both to physiology and environmental factors such as temperature and salinity. The isotopic composition of oxygen or carbon within carbonate skeletons is a function of isotopic composition of the seawater in which the skeleton was secreted as well as of water temperature. If temperature can be determined independently, the ratio of ^{18}O to ^{16}O (often expressed as a deviation from a standard and referred to as $\delta^{18}O$) can be used to determine whether a shell was secreted in water of normal (35%) or abnormal salinity. Conversely, if a given shell can be assumed, on independent evidence, to have come from a normal marine environment and is unaltered, then $\delta^{18}O$ may be used to determine paleotemperature. In general, carbonate secreted at lower temperatures is preferentially enriched with respect to ^{18}O, and so, $\delta^{18}O$ is useful for temperature determination.

Comparative Taphonomy

Taphonomy, which deals with processes and patterns of fossil preservation, has a critical dual role with respect to paleoecology. On the one hand, preservational processes impose distinct biases on the fossil record that must be considered carefully in any attempt at paleoecological reconstruction. On the other hand, the bodies and skeletons of dead organisms constitute biologically standardized sedimentary particles whose orientations, sorting, and general preservational condition bear the imprint of environmental processes active in the depositional environment. Comparative taphonomy uses the differential preservation of fossils as a source of paleoenvironmental information. The degree of preservation of fossils reflects biostratinomic processes, such as current-wave transport, decay, disarticulation, fragmentation and corrasion of skeletons, and fossil diagenetic factors acting after final entombment of the remains in sediment.

Evidence of mode of death of organisms may also provide critical details. For example, layers of beautifully preserved fish or reptile carcasses signify mass mortalities that involved changes in the water column itself. But such mass mortalities can be recorded only if they were also timed with burial events.

Soft tissues can be preserved only by exceptionally rapid burial in anoxic sediments followed by very early coating or impregnation by minerals. Such deposits not only yield important data on the paleobiology of organisms but also provide detailed insights into depositional environments.

Usually, however, only skeletal remains are preserved. Skeletons composed of bivalved shells (for example, brachiopods and pelecypods) or, particularly, of multiple articulated elements (for example, echinoderms, arthropods, and vertebrates) are sensitive indicators of episodic burial rates. Experimental studies have demonstrated the rapidity of disarticulation under normal marine conditions; most starfish, for example, disintegrate into ossicles in a few days. Hence, intact preservation of these organisms signals episodic burial events.

Individual skeletons, or parts of skeletons, may become physically fragmented, chipped, or abraded. Such evidence reflects the general degree of turbulence of a particular depositional environment. Similarly, the degree to which skeletal remains are size-or shape-sorted may signify the extent of current and wave processing. Skeletal destruction by bioerosion, physical abrasion, or chemical solution is generally a good indicator of residence time of skeletons on the sea floor prior to burial. The orientation of fossils may yield specific clues as to the extent and types of environmental energy. Pavements of convex-upward valves typically are associated with persistent current reworking, whereas abundant concave-upward shells may signify an episode of stirring of the shells from the sea bottom and resettlement during storms. Furthermore, alignment of elongated shells may provide data on the orientation of unidirectional currents or the propagation direction of oscillatory waves. Vertically embedded specimens of ammonoids are typical of water areas with depths less than 30 ft (10 m). Finally, the early diagenetic features of fossils reflected in solution, compaction, and mineralization may yield information about sediments and bottom water geochemistry, water pH and oxygen content.

Various aspects of biostratinomic and diagenetic fossil preservation can be combined to form predictive models of taphonomic facies or taphofacies. Certain suites of quantifiable preservational conditions, for example, characterize particular environments, and so their recognition by paleoecologists may help to "fingerprint" those environments.

Paleoautecology

Paleoautecology, the interpretation of modes of life (broadly, niches) of ancient organisms, involves a multidisciplinary approach. Although ancient modes of life cannot be determined completely, paleoecologists can often assign fossils to generalized guilds in terms of types of feeding, substrate preference, and degree of activity.

A thorough understanding of the biology of closest modern analogs is particularly important in any attempt to reconstruct paleoautecology. If the species or a closely related species is extant, then its mode of life, general physiology, and even behavior can be inferred with some confidence through the use of taxonomic uniformitarianism, provided that the biology of living relatives is well understood. "Living fossils," or relict extant taxa, such as Nautilus, sclerosponges, horseshoe crabs, and modern stalked crinoids provide valuable clues for interpreting the paleobiology of extinct organisms.

Functional Morphology

For extinct organisms that have no adequate modern analogs, alternative approaches, particularly functional morphology, provide some hints as to life modes. Comparative morphology seeks analogies between the anatomical features of fossil skeletons and those in living forms for which the function can be determined. In some cases, structures in unrelated organisms have evolved convergently, and their function may be interpreted by analogy. When no biological analog exists, a physical or mechanical model, or paradigm, may provide clues to interpreting structures in extinct organisms.

An experimental approach to functional morphology may also provide useful insights. Models of ammonite shells, for example, have been tested in flumes in which artificial currents are produced to determine frictional drag effects of shell shape and sculpture. Certain shell shapes were found to be more hydrodynamically streamlined and probably correspond to more rapidly swimming modes of life. Testing the resistance of different brachiopod shell architectures to crushing, as by predators, has shown that certain features of shell architecture, such as ribbing and deflections of shell margins, can increase shell rigidity.

Fossil Data

Certain natural experiments also shed light on the paleobiology of extinct organisms. The fact that oysters encrusted the shells of living ammonites has enabled paleontologists to calculate the buoyancy compensation capabilities of those ammonoids. Encrustation and boring of cephalopod shells by bryozoans and barnacles that grew preferentially aligned toward currents has demonstrated a predominance of forward swimming motion in these extinct hosts.

Remnants of soft parts, muscle scars, gut contents, and associated trace fossils all provide information useful in the reconstruction of ancient ecological niches. Rare occurrences of rapidly buried fossils in unusual positions can be interpreted as original life positions. Unusual associations with substrates or other organisms also provide insights. Finally, the consistent association of poorly understood fossil species with other fossils whose modes of life are well known or with sediments that indicate particular environments may help to establish the habits and environmental ranges of extinct forms.

Population Studies

Certain properties of species, such as mortality patterns, birth rates, and numbers of individuals per age class, can be studied only at the level of populations. Despite the difficulties of studying fossil populations, it is still possible to make some inferences about population parameters. For example, the distribution of individuals of a particular species into different age or size classes may yield some indirect data on the age-frequency distribution of a population that can be used to construct crude mortality curves showing age-at-death relationships. Some species may display a high juvenile mortality, a feature typically associated with stressed environments and rather opportunistic species; others, in stable environments, may display delayed mortality of particular ecological importance is the population strategy of a given species of organisms. Two end-member conditions have been recognized: opportunistic species, sometimes termed r-selected forms, and equilibrium, or k selected, species. Opportunistic organisms are typically rather generalized in habit and habitat preferences, are commonly stress-adapted, and display exceedingly high rates of reproductive maturation and fecundity. Extremely opportunistic, or "weedy," species of this sort are recognizable in the fossil record by their widespread distribution and occasional presence in extremely dense, mono-specific populations on single beds of rock that are otherwise barren of fossils. Equilibrium species, on the other hand, tend to occur in moderate or small numbers in a narrow range of environments commonly associated with diverse assemblages of other species, such as in reef environments. The distinction between equilibrium and opportunistic mode of life may have important implications for understanding the distribution and evolutionary patterns of fossil taxa as well as for interpreting the stability of particular ancient environments.

Paleosynecology

The study of interrelationships within organism communities that coexist in time and space is known as synecology. At the most basic level of synecology are the interacting pairs of organisms that coexist in a particular environment. Paleosynecology also involves study of ancient community structure and dynamics.

Organism Interactions

Interactions range from tolerance to symbiosis, which involves highly dependent and coevolved species pairs. Although interactions are very difficult to determine with fossils, there are cases where strong clues are observed. In some instances, organism interaction may be very indirect. For example, the accumulation of shells on a sea floor may lead to colonization by other organisms that require hard substrates to encrust onto or bore into, a process referred to as positive taphonomic feedback. Conversely, armoring of muddy sea bottoms with shell debris will inhibit burrowing organisms (negative taphonomic feedback).

Shallow-burrowing nuculid bivalves convert muddy sea bottoms into a water-rich pelleted floc. The high turbidity and instability of this fluid substrate may inhibit the settlement of many epifaunal suspension-feeding organisms. Such negative feedback is referred to as trophic group amensalism.

Mutualism, involving symbiotic algae called zooxanthellae, is inferred for many fossil reef-dwelling organisms, including various corals, sponges, and even some bivalves, based on taxonomic uniformitarianism as well as morphology and evidence for prolific skeletal growth. Such mutualism is difficult to substantiate for extinct groups, although distinctive patterns of carbon isotopes within skeletal carbonates may prove a useful fingerprint of secretion aided by zooxanthellae.

Many marine organisms use the skeletons of other living organisms as substrate or to obtain an elevated feeding position without having any effect on the hosts. Evidence of this type of commensal interaction is abundant in the fossil record.

Parasitic interactions are very difficult to observe in fossils because they normally involved the soft tissues of the host. However, rare evidence for paleopathology (fossil diseases) can be documented from Paleozoic times onward in certain organisms such as echinoderms or vertebrates that have an internal skeleton, or endoskeleton. For example, malformations in fossil crinoids may record parasitism.

Fossil evidence for competition is best seen in cases of spatial competition. For example, certain types of bryozoans appear to overgrow other species preferentially. Many aspects of fossil distribution have been attributed to the effects of competition or the evolutionary response for reducing competition by niche partitioning. Examples include the subdivision of many marine communities into distinct feeding groups based on vertical height (tiering) above and below the sediment-water interface. Some researchers have claimed that competition is a primary motor of evolutionary change, often alluding to Darwin's analogy of the wedge, in which more and more species are packed into a particular ecospace by increasingly finely divided specialized niches.

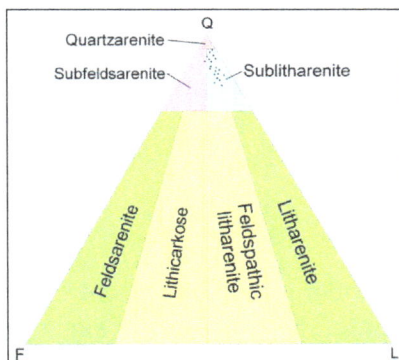

Classification of shales in terms of percentages of three benthic habitat groups epifaunal suspension-feeding bivalves (those living at the sediment-water interface and filtering seawater), infaunal (living within and feeding on the sediment), and suspension-feeding groups.

Predation or carnivory is probably one of the most significant ecological interactions in any environment. Direct predatorprey links are difficult to establish in the fossil record, but there are several lines of evidence that may be used. Bite marks of particular types, such as tooth marks of mosasaurs on ammonoids, provide one line of evidence, so do boreholes of predatory snails on particular prey species and remnants of prey shells preserved in the stomach contents or coprolites (fossilized feces). In turn, numerous morphological trends may signify antipredatory adaptations. The fossil record of predation extends back to the Early Cambrian, as evidenced by bite marks in trilobites, which commonly show a preference for the right side of the prey. The frequency of healed and unhealed predatory fractures in some shells increases significantly in the Paleozoic in concert with the rise of fossil evidence for shell-crushing predators. The earliest shell-drilling snails appear to have been Ordivician in age, but the habit of drilling shells for predation probably evolved independently at least four times in different groups of gastropods.

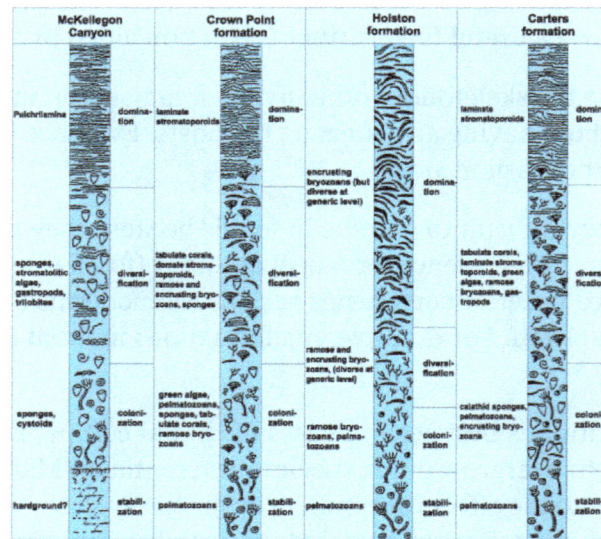

Comparison of four developmental stages in four ancient reef masses.

Paleocommunities

The fossil record contains highly biased remnants of past communities or paleocommunities. Paleocommunities are generally recognized as recurring associations of fossil species. Multivariate statistical techniques such as cluster analysis and ordination analysis are commonly employed to aid in discerning the recurrent groupings of fossil species, or persistent gradients of species composition. Such analyses are based upon field studies in which data on the presence, absence, or relative abundance of fossil taxa have been recorded in a large number of samples, typically from many stratigraphic levels.

Taphonomic Biases

Statistically defined groupings may or may not represent real ecological entities. For example, in most offshore marine environments, the transport of skeletons between environments is minimal. However, because of differential preservation, the proportions of organisms in the living assemblages (biocoenoses) are not always faithfully reproduced in the death assemblages of skeletal remains (taphocoenoses). Nearly all soft-bodied organisms are lacking in the death assemblage, and those with fragile skeletons tend to be underrepresented. Moreover, because of the accumulation of skeletons over extended periods of time, death assemblages commonly display mixtures of organisms that inhabited slightly differing environments at different times, a phenomenon referred to as time averaging. Fossil assemblages actually may be more diverse than living assemblages of skeletonized organisms at any one time. They record a very biased and averaged-out view of communities that existed over a long period of time.

Relationship to Environments

In most studies of paleocommunities, recurrent groups can be related to environments, as inferred from independent evidence such as rock type, sedimentary structures, taphonomy, trace elements, and isotopic studies. Classic studies modeled paleocommunity distribution patterns on relative bathymetry or distance from shoreline, but many later studies emphasized the control of paleocommunity distribution by multiple factors. Depth-related factors such as turbulence, light penetration, and oxygen level are clearly important controls in many cases. However, sedimentation-related factors such as rates of deposition, turbidity, and substrate consistency may be equally important, giving rise to a much more complex array of paleocommunities.

After recurring associations of fossils are recognized, they are generally analyzed in terms of organism interactions and trophic (feeding) relationships. Primary producers of ancient ecosystems, such as algae, are likely to be poorly preserved or absent from the fossil record, but some links in ancient food chains may be recognizable. One aspect of paleocommunity structure that is commonly analyzed is the proportion of different feeding (trophic) and life-habit guilds. Certain marine paleocommunities are dominated by skeletons of infaunal deposit feeders, others by epifaunal suspension feeders. Unfortunately, live-dead comparisons in modern communities suggest that the original proportions of various life habitat groups are not preserved in the fossil record. But the biased trophic proportions of the taphocoenoses may still relate in a meaningful way to the original environment. Consistent differences in the proportions of infaunal suspensionfeeding, infaunal deposit-feeding, and epifaunal suspension-feeding bivalves have been detected in differing ancient oxygen-restricted facies.

Temporal Changes

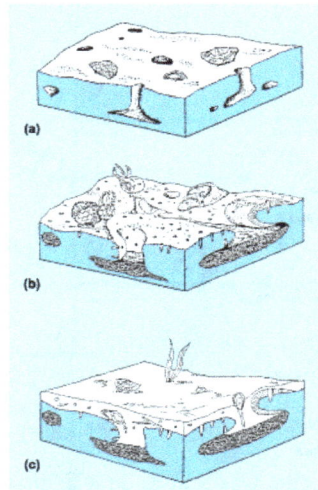

Formation and burial sequence of Bobcaygeon hardground

In figure, (a) Soft-bottom community of strophomenid brachiopods and infaunal burrowers inhabiting carbonate mud. (b) Hardground community consisting of boring and encrusting organisms. (c) Post-hardground community inhabiting muds that blanketed the hardground.

Communities and paleocommunities are not static entities in time, but undergo important structural changes on at least three different time scales: succession, replacement, and evolution. Because it operates on a very short time scale, from decades to centuries, ecological succession can be resolved only in a few fossil samples. Some instances of supposed ecological succession, such as encrusting communities upon shells, may in fact reflect taphonomic feedback. Allogenic succession represents changes in communities induced by physical environmental change. Good examples are seen in many hardgrounds, areas of early lithified sea floors.

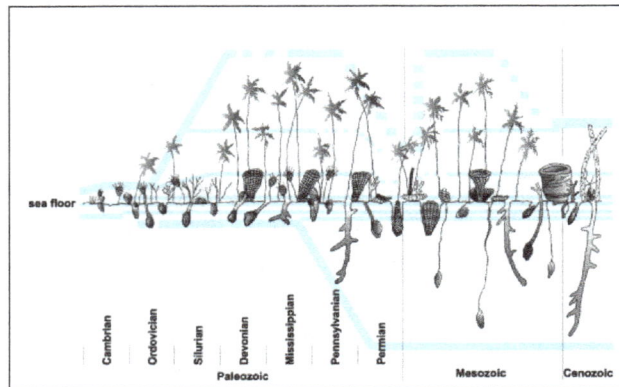

Tiering or vertical depth stratification in marine communities through time.

The rapid rise of the highest-tier (level) organisms (crinoids or sea lilies, bryozoans) from a few centimeters to over a meter above the sea floor during the Ordovician Period. Branching bryozoans and shorter-stemmed crinoids took advantage of an intermediate tier 10–20 cm above the sea bottom, while burrowing clams and worms dug down to a tier about the same distance into the sediment. In the middle Paleozoic, still much deeper burrowing forms evolved the ability to mine sediments down to nearly a meter.

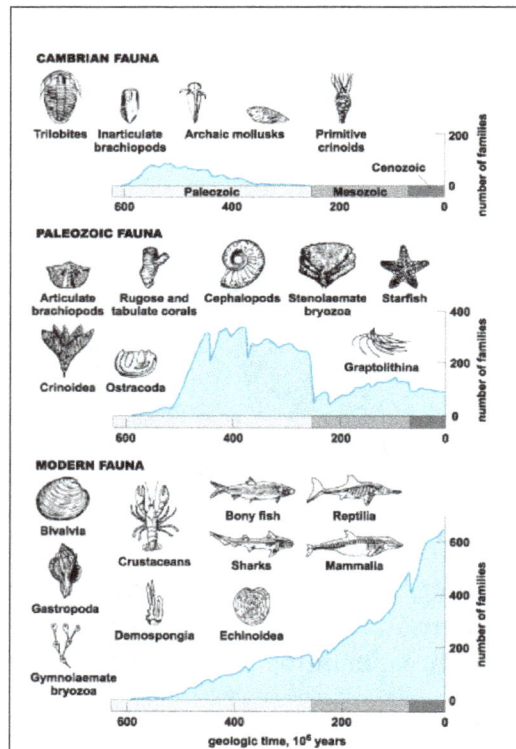

The three great marine evolutionary faunas of the Phanerozoic

The Cambrian fauna, composed of trilobites, primitive groups of brachiopods, and mollusks arose early during that period to a diversity of about 50 families, then dwindled during the later Paleozoic as more archaic groups migrated offshore and were replaced by the "Paleozoic fauna," typified by rugose and tabulate corals, brachiopods, bryozoans, crinoids, and graptolites. The latter diversified rapidly in the Ordovician Period to over 300 families and then fluctuated around this level until the end of the Paleozoic Era. The great Permian-Triassic extinction reduced the "Paleozoic fauna" and may have favored the rise of the "Modern fauna" during the Mesozoic and Cenozoic eras with diversities as high as 620 families, including especially mollusks, crustaceans, and both sharks and bony fishes.

Longer-term changes in community composition, encompassing thousands of years, are not truly succession, but instead record allogenic effects such as sea level or climate variations. These changes are properly termed community replacement, and involve wholesale migration or restructuring of communities at particular locations due to changing environments. In many instances, particular fossil assemblages appear to track shifts in preferred environments and facies within sedimentary cycles. Habitat tracking may provide important clues to deciphering patterns of environmental fluctuations such as transgressive-regressive cycles. On a scale of millions of years, communities show evolutionary changes because their component species have evolved.

Larval Ecology and Evolution

Ecological patterns such as larval type affect overall patterns in life history. Larval ecology of marine animals controls their geographic distribution. Species with long-lived larvae may be dispersed much more widely than forms with shortlived planktonic phases or direct development from eggs. In turn, geographic distribution, whether localized or cosmopolitan (global), undoubtedly plays an important role in their tendency toward speciation as well as extinction. Thus, it may be possible to develop models to better explain evolutionary patterns in different groups in relation to paleoecology.

Evolutionary Paleoecology

Organisms evolve within the context of other organisms, not in a vacuum. There is substantial fossil evidence to indicate increasing complexity of organism interactions through time. This escalation in the intensity of predatory interactions, for example, may have important implications for evolutionary change. For example, trends of increased spinosity, greater shell thickness, increasingly restricted apertures, and other antipredation adaptations may reflect the intensification of predatory behavior by shell-boring and crushing predators.

Increased vertical stratification or tiering in marine-level bottom communities through time has been recognized. Cambrian communities possessed mainly low-lying suspension-feeding and scavenging organisms that lived mostly just above or below the sediment-water interface. By mid-Paleozoic time, crinoids extended up to heights of several feet or more off the sea floor, and various burrowers extended downward a couple of feet or more into the sediment. The Mesozoic rise of deep-burrowing clams and other infauna increased the infaunal tier to over 3 ft (1 m). The increased vertical structuring of these communities may represent a response to increasing crowding. By feeding at different levels in the water and substrate, organisms were able to further subdivide the resources of a given environment.

Marine animals form a hierarchy of ecological units through the Phanerozoic time interval. These range from blocks of relative stability at time scales of a few million years, to broader intervals of general stability of faunas, to three great evolutionary faunas.

First, at a scale of a few million years, groups of species may show considerable ecological stability punctuated by episodes of abrupt change. Brett and Baird introduced the concept of "coordinated stasis" to describe a pattern of approximately concurrent long-term stability and abrupt change in many taxa. During a large proportion of geologic time a majority of genera and, in some cases, species show little or no change in morphology. Moreover, general groups of communities or "biofacies" also may be similar throughout blocks of stability referred to as "ecological-evolutionary units and subunits." These relatively stable intervals, spanning up to several million years, are punctuated by much shorter intervals, perhaps a few tens of thousands of years, of abrupt change across many biofacies, local extinction of many longstanding lineages, immigration and emigration from the local basin, and general faunal turnover. The original example of the Silurian–Devonian (380 to 440 million year old) fossil assemblages of eastern North America—in particular, the Middle Devonian Hamilton Group—features examples of assemblages, separated by up to 5 million years, with nearly identical composition and similar guild structure and even relative abundance. Consideration of a larger number of case studies ranging in age from Cambrian to modern suggests that this original example represents one end member in an array of conditions ranging from similar cases but some with somewhat more species level variability, to examples of nearly continual change in species composition, and biofacies ecological structure. This variability probably depends on local environmental variability. The observed pattern of similarities between samples from cases of coordinated stasis could imply a form of stable, lock-step tracking of certain well-organized "communities." However, this pattern could also be the result of recurrence of a similar assemblage due to persistence of environmental gradients and because species do not drastically change their habitat preferences through time. The retention of habitat preferences by species is perhaps the most important aspect of ecological stasis. It would appear that under appropriate conditions species can simply track shifting preferred environments for millions of years rather than adapt to local change.

At the next larger level, marine communities appear to show strong similarities of family- and genus-level composition, as well as general ecological structure (guilds, trophic structure, diversity, and so on) for tens of millions of years. These blocks of relative stability, termed ecological-evolutionary units (EEUs) by Boucot, were terminated by major extinctions. Raup and Sepkoski also recognized five major mass extinctions—the "Big Five" (Late Ordovician, Late Devonian, Permian-Triassic, Late Triassic, and Cretaceous-Tertiary) that stand out from background rates of extinction. These and lesser mass extinctions played critical roles in restructuring the ecology of the biosphere, including changes in guild structure and tiering patterns. Ecological-evolutionary units are bounded by major biotic turnover events, involving widespread extinctions including the "Big Five" mass extinctions. Again, the EEU concept implies that the ecological history of life was not one of continuous, gradual change. Rather, it was characterized by extended periods of near equilibrium that were interrupted by much shorter periods of crisis and major ecological restructuring.

The largest scale of faunal pattern consists of "evolutionary faunas." By analyzing patterns of marine family and genus level diversity using a large database, Sepkoski recognized three such units through the past 540 million years of the Phanerozoic Eon, each characterized by a different pattern or trajectory of diversification. The "Cambrian fauna"—typified by trilobites, lingulid brachiopods, and certain primitive groups of mollusks and echinoderms—appeared during the earliest Paleozoic, diversified in the Cambrian, and then began to decline as the second or "Paleozoic fauna" diversified. The latter was characterized by rugose and tabulate corals, articulate brachiopods, bryozoans, and crinoids which formed the major faunas of shallow seas from the Ordovician to the Permian Period and displayed a relatively stable "platform" of family diversity. Finally, the "Modern fauna," characterized by mollusks and crustaceans, arose in nearshore environments during the early Paleozoic, but expanded greatly following the end Permian mass extinctions. Sepkoski and Sheehan recognized that evolutionary innovations tended to arise first in shallow, nearshore environments. Through time the newly arising groups typical of the "Paleozoic" and then the "Modern" faunas tended to spread offshore, while more archaic forms were displaced to deep ocean "refugia." This is one of the most profound of all paleoecological patterns, and the explanation of this pattern remains imperfectly understood. It may imply that stressed nearshore settings favor evolution of new life strategies and/or that there has been a general intensification of energy utilization through time such that more archaic "Cambrian" or "Paleozoic" faunas were relatively "low energy" and had less competitive ability than physiologically more sophisticated, "high-energy" Modern faunas.

References

- Concepts-of-ecology-structural-and-functional-concept-of-ecology, ecology: yourarticlelibrary.com, Retrieved 25 May, 2019

- Ecological-disturbance, science: britannica.com, Retrieved 14 April, 2019

- What-is-ecology: environment-ecology.com, Retrieved 2 March, 2019

- Habitats-basics: thoughtco.com, Retrieved 7 August, 2019

- System-ecology: systemsacademy.io, Retrieved 13 May, 2019

- Human-ecology: gerrymarten.com, Retrieved 11 February, 2019

- Population-ecology: nature.com, Retrieved 18 January, 2019

- Popecology, desert-agriculture: bgu.ac.il, Retrieved 9 June, 2019

- Plant-ecology, earth-and-planetary-sciences: sciencedirect.com, Retrieved 10 January, 2019

- Ecology-and-habitats, animal: britannica.com, Retrieved 8 February, 2019

- Applied-Ecology: researchgate.net, Retrieved 10 April, 2019

- Paleoecology-from-Access-Science: creighton.edu, Retrieved 27 July, 2019

Chapter 2

Understanding Ecosystems

The community of living organisms along with the non-living components in their surroundings, which interact as a system is known as an ecosystem. There are several types of ecosystems such as mountain ecosystem, desert ecosystem and aquatic ecosystem. The chapter closely examines the key aspects of these ecosystems along with their components to provide an extensive understanding of the subject.

An ecosystem is a geographic area where plants, animals, and other organisms, as well as weather and landscape, work together to form a bubble of life. Ecosystems contain biotic or living, parts, as well as abiotic factors, or nonliving parts. Biotic factors include plants, animals, and other organisms. Abiotic factors include rocks, temperature, and humidity.

Every factor in an ecosystem depends on every other factor, either directly or indirectly. A change in the temperature of an ecosystem will often affect what plants will grow there, for instance. Animals that depend on plants for food and shelter will have to adapt to the changes, move to another ecosystem, or perish.

Ecosystems can be very large or very small. Tide pools, the ponds left by the ocean as the tide goes out, are complete, tiny ecosystems. Tide pools contain seaweed, a kind of algae, which uses photosynthesis to create food. Herbivores such as abalone eat the seaweed. Carnivores such as sea stars eat other animals in the tide pool, such as clams or mussels. Tide pools depend on the changing level of ocean water. Some organisms, such as seaweed, thrive in an aquatic environment, when the tide is in and the pool is full. Other organisms, such as hermit crabs, cannot live underwater and depend on the shallow pools left by low tides. In this way, the biotic parts of the ecosystem depend on abiotic factors.

The whole surface of Earth is a series of connected ecosystems. Ecosystems are often connected in a larger biome. Biomes are large sections of land, sea, or atmosphere. Forests, ponds, reefs, and tundra are all types of biomes, for example. They're organized very generally, based on the types of plants and animals that live in them. Within each forest, each pond, each reef, or each section of tundra, you'll find many different ecosystems.

The biome of the Sahara Desert, for instance, includes a wide variety of ecosystems. The arid climate and hot weather characterize the biome. Within the Sahara are oasis ecosystems, which have date palm trees, freshwater, and animals such as crocodiles. The Sahara also has dune ecosystems, with the changing landscape determined by the wind. Organisms in these ecosystems, such as snakes or scorpions, must be able to survive in sand dunes for long periods of time. The Sahara even includes a marine environment, where the Atlantic Ocean creates cool fogs on the Northwest African coast. Shrubs and animals that feed on small trees, such as goats, live in this Sahara ecosystem.

Even similar sounding biomes could have completely different ecosystems. The biome of the Sahara Desert, for instance, is very different from the biome of the Gobi Desert in Mongolia and China.

The Gobi is a cold desert, with frequent snowfall and freezing temperatures. Unlike the Sahara, the Gobi has ecosystems based not in sand, but kilo meters of bare rock. Some grasses are able to grow in the cold, dry climate. As a result, these Gobi ecosystems have grazing animals such as gazelles and even taking, an endangered species of wild horse.

Even the cold desert ecosystems of the Gobi are distinct from the freezing desert ecosystems of Antarctica. Antarctica's thick ice sheet covers a continent made almost entirely of dry, bare rock. Only a few mosses grow in this desert ecosystem, supporting only a few birds, such as skuas.

Threats to Ecosystems

For thousands of years, people have interacted with ecosystems. Many cultures developed around nearby ecosystems. Many Native American tribes of North Americas Great Plains developed a complex lifestyle based on the native plants and animals of plains ecosystems, for instance. Bison, a large grazing animal native to the Great Plains, became the most important biotic factor in many Plains Indians cultures, such as the Lakota or Kiowa. Bison are sometimes mistakenly called buffalo. These tribes used buffalo hides for shelter and clothing, buffalo meat for food, and buffalo horn for tools. The tallgrass prairie of the Great Plains supported bison herds, which tribes followed throughout the year.

As human populations have grown, however, people have overtaken many ecosystems. The tallgrass prairie of the Great Plains, for instance, became farmland. As the ecosystem shrunk, fewer bison could survive. Today, a few herds survive in protected ecosystems such as Yellowstone National Park.

In the tropical rain forest ecosystems surrounding the Amazon River in South America, a similar situation is taking place. The Amazon rain forest includes hundreds of ecosystems, including canopies, understories, and forest floors. These ecosystems support vast food webs.

Canopies are ecosystems at the top of the rainforest, where tall, thin trees such as figs grow in search of sunlight. Canopy ecosystems also include other plants, called epiphytes, which grow directly on branches. Under story ecosystems exist under the canopy. They are darker and more humid than canopies. Animals such as monkeys live in understory ecosystems, eating fruits from trees as well as smaller animals like beetles. Forest floor ecosystems support a wide variety of flowers, which are fed on by insects like butterflies. Butterflies, in turn, provide food for animals such as spiders in forest floor ecosystems.

Human activity threatens all these rain forest ecosystems in the Amazon. Thousands of acres of land are cleared for farmland, housing, and industry. Countries of the Amazon rain forest, such as Brazil, Venezuela, and Ecuador, are underdeveloped. Cutting down trees to make room for crops such as soy and corn benefits many poor farmers. These resources give them a reliable source of income and food.

However, the destruction of rain forest ecosystems has its costs. Many modern medicines have been developed from rain forest plants. Curare, a muscle relaxant, and quinine, used to treat malaria, are just two of these medicines. Many scientists worry that destroying the rain forest ecosystem may prevent more medicines from being developed.

The rain forest ecosystems also make poor farmland. Unlike the rich soils of the Great Plains, where people destroyed the tallgrass prairie ecosystem, Amazon rain forest soil is thin and has few nutrients. Only a few seasons of crops may grow before all the nutrients are absorbed. The farmer or agribusiness must move on to the next patch of land, leaving an empty ecosystem behind.

Rebounding Ecosystems

Ecosystems can recover from destruction, however. The delicate coral reef ecosystems in the South Pacific are at risk due to rising ocean temperatures and decreased salinity. Corals bleach, or lose their bright colors, in water that is too warm. They die in water that is not salty enough. Without the reef structure, the ecosystem collapses. Organisms such as algae, plants such as seagrass, and animals such as fish, snakes, and shrimp disappear.

Most coral reef ecosystems will bounce back from collapse. As ocean temperature cools and retains more salt, the brightly colored corals return. Slowly, they build reefs. Algae, plants, and animals also return. Individual people, cultures, and governments are working to preserve ecosystems that are important to them. The government of Ecuador, for instance, recognizes ecosystem rights in the countrys constitution. The so-called Rights of Nature says Nature or *Pachamama* [Earth], where life is reproduced and exists, has the right to exist, persist, maintain and regenerate its vital cycles, structure, functions and its processes in evolution. Every person, people, community or nationality, will be able to demand the recognitions of rights for nature before the public bodies. Ecuador is home not only to rain forest ecosystems, but also river ecosystems and the remarkable ecosystems on the Galapagos Islands.

Tall grasses and Bison bison—must be the tallgrass prairie ecosystem.

Desert Ecosystem

A desert ecosystem is a community of organisms that live together in an environment that seems to be deserted wasteland.

Desert ecosystems can be hot (as in the sandy Sahara) or cold (as on the peaks of mountains where the high altitude makes conditions very harsh) but both hot and cold deserts have in common the fact that they are difficult for organisms to inhabit.

A desert ecosystem is generally witnesses little rainfall, resulting in less vegetation than in more humid areas of the globe. Look closely at any seemingly deserted piece of land and you will usually be able to see:

- Numerous insects living in communities.

- An abundance of plant life.

- Mammals and birds.

- In addition, micro-organisms such as bacteria will also be present in this ecosystem, though they are not visible to the naked human eye.

In desert ecosystems, the plant and animal life that lives there will have evolved so that they can combat the harsh conditions (for example, they will have evolved to store water supplies in their bodies as water is very scarce in deserts).

In general, deserts are made up of a number of abiotic components – including sand, the lack of moisture, and hot temperatures – basically anything that makes up an ecosystem that isn't alive. However, there are also a number of biotic factors that affect deserts, which include living things, such as plants and animals.

Abiotic Components

Temperate Deserts

Antarctica is an example of a temperate desert. The temperatures are actually so cold, they could lead to the death of humans. In order to survive, the animals that live in these kinds of deserts have adapted with the passage of time. The ways they have done this is by adding extra layers of fat, or needing less food and energy in order to survive.

Subtropical Deserts

These deserts are too hot for many plants and animals to handle. The animals who call these deserts home have adapted to having less water. Because it is so hot during the day, they have become nocturnal, getting out during the night when it is cooler and easier to manoeuvre without getting overheated. But, because the nights are cold, they have had to become accustomed to the colder nights. Plants have had to adjust to having less water, so they are sparse and often close to the ground.

Location

Mountains-There are two major factors in the deserts' creation; mountains' rain shadows and the large circulation of global winds. As water-filled air is pushed up the mountain slopes, it cools then drops water on that particular side of the mountain. In the event of larger mountain ranges, very little water makes it to the other side. Therefore deserts are often found near mountainous areas, such as:

- The Caucasus Mountains in Asia, where the Kara kum and Kyzyl Kum deserts are.

- The Atacama Desert, which is partly caused by the Andes Mountains in Chile.

- Parts of California, where the Santa Cruz mountains are.

- The Sahara desert, which is affected by a number of different mountain ranges.

Wind patterns- Global wind patterns, which are complicated, play a significant role in where deserts are located. Winds that circle the globe are the result of the difference between warmer equatorial temperatures as well as the polar temperatures that are cooler. After air has been warmed at the equator, it moves upward. Then it moves toward the north pole and toward the south pole, where it loses moisture, cools off and then sinks before returning to the equator. Therefore, stable wind patterns and shifting global patterns can contribute to where a desert is.

The passage of time greatly influences where and how deserts form. As time has passed, the locations of deserts have moved through the passage of geologic time. This change has been the result of the uplifting of mountain ranges and the continental drift. The horse latitudes are where more deserts are situated, which is generally straddling the Tropic of Capricorn and the Tropic of Cancer, which falls between 15 and 30 degrees to the equator's north.

There are geologically ancient deserts, such as the Sahara Desert in northern Africa, which is 65 million years old or the Kalahari in central Africa. In North America, three of the four major deserts are within a geological region called the Range Province and the Basin, which falls between the Sierra Nevadas and the Rocky Mountains then extending into the state of Sonora in Mexico.

The forces of erosion thousands of years past shaped the desert landscapes during heavy rainfall. The rocky mountain slopes and hillsides caught the rain, which picked up loose sediment, sand, cobbles and boulders then moved them. As gravity caused the water to be carried downhill, sediment was moved down to the basin. At the bottom of the mountain, the water spread out across a broad area where the mouths of canyons were widened.

Temperature

The temperature of a given desert will vary due to its geographic location. However, a characteristic of all deserts is the dryness. Heat is reflected by water vapor, which is either in the form of cloud cover or humidity, resulting in a cooling effect. Because of the reactions and the characteristics, deserts experience extreme temperatures, regardless of whether it is heat or cold.

The temperature fluctuations can result in other effects. Cool air sinks and warm air rises, so the fast changes of temperatures cause the air to move fast from one place to another. Because of that, deserts are windy, and those conditions contribute to evaporation. About 90% of available sunlight is transmitted by clear dry air, which in comparison to a typical humid climate seeing 40% of the available sunlight. The additional sunlight has ultraviolet radiation, which can cause major damage to plants, animals and people.

Precipitation

The desert environment has an unpredictable and uneven of the precipitation that is does receive, although that precipitation is minimal in nature. Precipitation amounts can vary from year to year. Some years it may seem as though the desert has gotten more rainfall than usual, but most years have very little rainfall. There can actually be entire years that the desert doesn't see a drop of rain.

Biotic Components

Plant Life

Water is important everywhere and for every living thing. And it is, of course, extremely important in the desert. Because of the lack of water, the plants have made major adaptations.

Plant Adaptations

- The seeds of annual plants stay dormant until a time when there is adequate rainfall available to support a young plant.

- Cacti and other succulent plants store water in their spines, which are residual leaves. The stem is where photosynthesis takes place and the stem has pleats that are able to expand fast when rain falls.

- Evergreens have way cuticles and sunken stomata on shrubs that help hold water and prevent it from escaping. As an example, the holly plant's leaves are held at 70-degree angles so the sun only hits its sides. When the sun sinks low in the sky, the entire leaf is exposed. A fine salt covering is on the leaves and that helps reflect the sun off of the plant.

More than a fifth of the earth's land is comprised of deserts. The lack of water can create a survival problem for any humans, animals, plants or organisms. Besides the low rainfall, deserts experience a high amount of water loss from evaporation from the ground and through transpiration of plants. Evapotranspiration is from the combination of evaporation and transpiration. Potential evapotranspiration is how much water that would be lost by transpiration and evaporation if they were possible. Scientists measure this amount under controlled conditions with a large pan of water.

Soil in the desert is known for its coarseness, which permits the little moisture that is in it to pass through quickly, which means it is not as available for plants. Salts accumulate as a result from the high evaporation rate. The soil becomes alkaline and limits plant growth, which is also known as primary productivity.

Animal Life

Because of the entire process required to maintain life in the desert, the impact is that the size of individual animals is limited as well as the size of animal populations. The extremes of heat and aridity result in deserts being one of the most fragile of the ecosystems in the world.

Visitors to the desert should also take the proper precautions to protect themselves as the environment is much different than any other location.

Despite common beliefs that things can't live in the desert there are a number of creatures that have learned to survive on the distinctive plant life and in the difficult conditions.

- Large mammals like camels make their homes in the desert, and are suited to travel long periods of time without water. Lions live in the deserts of Africa, although they are endangered due to changing weather patterns and the presence of humans.

- Small rodents find homes in the desert, with variations from gerbils to hedgehogs. Larger hyenas and jackals are also often found in deserts.

- Lizards and snakes are particularly suited to the dry, hot climate of the desert, as are amphibious creatures like a number of toads and salamanders.

Mountain Ecosystem

Mountain ecosystem is complex of living organisms in mountainous areas. Mountain lands provide a scattered but diverse array of habitats in which a large range of plants and animals can be found. At higher altitudes harsh environmental conditions generally prevail, and a treeless alpine vegetation, upon which the present account is focused, is supported. Lower slopes commonly are covered by montage forests. At even lower levels mountain lands grade into other types of landform and vegetation—e.g., tropical or temperate forest, savanna, scrubland, desert, or tundra.

The largest and highest area of mountain lands occurs in the Himalaya-Tibet region; the longest nearly continuous mountain range is that along the west coast of the Americas from Alaska in the north to Chile in the south. Other particularly significant areas of mountain lands include those in Europe (Alps, Pyrenees), Asia (Caucasus, Urals), New Guinea, New Zealand, and East Africa.

Viewed against a geologic time frame, the processes of mountain uplift and erosion occur relatively quickly, and high mountain ranges therefore are somewhat transient features. Many mountains are isolated from other regions of similar environmental conditions, their summit regions resembling recently formed islands of cool climate settled amid large areas of different, warmer climates. Because of this isolation, mountaintops harbour a distinct biota of youthful assemblages of plants and animals adapted to cold temperatures. At lower elevations, however, some mountains are able to provide refuges for more ancient biota displaced by environmental changes. Also, mountainous vegetation usually has been affected less by human activities than the surrounding areas and so may harbour plants and animals that have been driven out by anthropogenic disturbances that have occurred elsewhere.

During the glacial intervals of the past two million years—the Ice Ages of the Northern Hemisphere—habitats suitable to cold-adapted biota covered much larger areas than they do today, and considerable migration of cold-adapted plants and animals occurred. Arctic biota spread south across large areas beyond the greatly expanded ice sheets that covered much of northern North

America, Europe, and Asia. When climatic conditions ameliorated, these organisms retreated both northward toward Arctic latitudes and uphill into areas of mountainous terrain. This history explains, for example, the close similarities between the fauna and flora of high mountains such as the European Alps and the Arctic far to their north.

In the tropics, however, little opportunity for similar overland movement of cold-adapted biota was possible because vast forestland in the tropical lowlands formed a barrier to migration. The organisms therefore have been isolated more completely from those of other cold environments. Despite this situation, colonization of tropical high mountains has occurred. Birds are particularly mobile, and some of temperate affinity found their way to equatorial peaks; for example, in the mountains of New Guinea are found pipits and thrushes that have no near relatives in the adjacent tropical lowlands. Migrating birds may have been the vectors for the seeds of cold-adapted plants growing in the same places, which also lack tropical lowland relatives.

Populations of mountain species are commonly both small—although fluctuating—and isolated and often have evolved over a relatively short period of time. It is therefore not unusual to encounter related but distinct species on separate mountain peaks. This recent and rapid production of new species contributes significantly to the biodiversity and biological importance of mountain lands.

Environment

Mountain environments have different climates from the surrounding lowlands, and hence the vegetation differs as well. The differences in climate result from two principal causes: altitude and relief. Altitude affects climate because atmospheric temperature drops with increasing altitude by about 0.5 to 0.6 °C (0.9 to 1.1 °F) per 100 metres (328 feet). The relief of mountains affects climate because they stand in the path of wind systems and force air to rise over them. As the air rises it cools, leading to higher precipitation on windward mountain slopes (orographic precipitation); as it descends leeward slopes it becomes warmer and relative humidity falls, reducing the likelihood of precipitation and creating areas of drier climate (rain shadows).

While these general principles apply to all mountains, particular mountain climates vary. For instance, mountains in desert regions receive little rain because the air is almost always too dry to permit precipitation under any conditions—e.g., the Ahaggar Mountains in southern Algeria in the middle of the Sahara. Latitude also can affect mountain climates. On mountains in equatorial regions winter and summer are non-existent, although temperatures at high altitude are low. Above about 3,500 metres frost may form any night of the year, but in the middle of every day temperatures warm substantially beneath the nearly vertical tropical sun, thus producing a local climate of "winter every night and spring every day." For example, at an altitude of 4,760 metres in Peru, temperatures range from an average minimum of about −2 °C (28 °F) to average maximum values of 5 to 8 °C (41 to 46 °F) in every month of the year.

By contrast, mountains at temperate latitudes have strongly marked seasons. Above the tree line during the summer season, temperatures high enough for plant growth occur for only about 100 days, but this period may be virtually frost-free even at night. During the long winter, however, temperatures may remain below freezing day and night. Snow accumulation and the phenomena this type of precipitation may cause, such as avalanching, are important ecological factors in temperate but not tropical mountain regions.

Microclimate variations are also important in mountain regions, with different aspects of steep slopes exhibiting contrasting conditions due to variations in precipitation and solar energy receipt. In temperate regions mountain slopes facing the Equator—southward in the Northern Hemisphere and northward in the Southern Hemisphere—are significantly warmer than opposite slopes. This can directly and indirectly influence the vegetation; the length of time snow remains on the ground into spring affects when vegetation will emerge, and this in turn affects the land's utility for grazing. Even in the tropics, aspect-related climate and vegetation contrasts occur, in spite of the midday vertical position of the sun. In New Guinea, for example, slopes facing east are warmer and drier and support certain plants at higher altitudes than slopes facing west, because the prevailing pattern of clear, sunny mornings and cloudy afternoons affects the amount of solar energy received by these contrasting aspects.

Mountain soils are usually shallow at higher altitudes, partly because the soil has been scraped off by the ice caps that formed on most high mountains throughout the world during the last glacial interval that ended about 10,000 years ago. Soils are generally poor in nutrients important to plants, especially nitrogen. Rapid erosion of loose materials is also common and is exacerbated by frost heaving, steep slopes, and, in temperate regions, substantial runoff of meltwater in spring. Soil is virtually absent on rocky peaks and ridges. However, because of the cool, wet climate, many mountain areas accumulate peat, which creates local deep, wet, acidic soils. In volcanic regions tephra (erupted ash) may also contribute to soil depth and fertility.

Considering the wide geographic extent of mountains and their resultant geologic and climatic variability, it is remarkable that they exhibit such a clear overall pattern in vegetation. The major structural feature of vegetation on mountains in all regions—except in very dry or very cold places—is tree line. (This characteristic is sometimes called timberline or forest limit, although strictly speaking the former term refers to the uppermost reaches that commercial-size timber trees attain and the latter term refers to a closed forest.) Above a critical level, which may vary between slopes on the same mountain and which is much higher on mountains at lower latitude, the climate becomes too harsh to permit tree growth; beyond that level grows alpine vegetation, dominated by herbaceous plants, such as grasses and forbs, or by low shrubs.

In general, the altitude at which the tree line occurs is determined by that at which the mean temperature in the warmest month approximates 10 °C (50 °F), provided moisture is not a limiting factor. This is not precisely the case under all circumstances, however; for example, in some tropical regions that have a yearlong growing season, forests can grow in conditions slightly cooler than this. Nevertheless, the value holds true in most regions, especially in the temperate zones. It reflects a fundamental requirement for a sufficient level of photosynthesis to occur to support the growth of tree trunks.

A relatively narrow belt of intermediate or mixed vegetation—the subalpine—usually exists between the forests below and the alpine vegetation above. In the subalpine of temperate mountains, stunted, usually infertile individuals of various tree species survive, despite blasts of windblown snow, frost damage, and desiccation. These deformed shrub-size trees are called krummholz.

Although the overall pattern in which forest gives way to alpine vegetation is common to mountains at all latitudes, the factors responsible for it are not the same in all places. In temperate-zone mountains, the brevity of the growing season is of paramount importance because tree shoot tissues that have had insufficient time to harden before growth ceases and winter conditions begin may die when frozen. Other factors that damage or kill shoots or entire trees in winter in this region at temperate latitudes include the abrasion of buds by windblown snow crystals, desiccation of shoots just above the snowpack where they are exposed to direct and snow-reflected solar radiation—especially late in winter as the sun angle rises—and infection of shoots beneath the snow by snow fungus. Freezing injury to roots may also occur if the insulating layer of snow is blown from the ground surface.

In the tropics, these phenomena are not experienced. Snowfall is not restricted to a single winter season, and when it occurs it usually melts quickly. Snow therefore does not accumulate as a thick, continuous cover except at altitudes above the upper limit of most plant life. For example, in Venezuela the tree line lies below 4,000 metres, even where there has been no human disturbance, but virtually permanent snow patches are not encountered until about 5,000 metres, where no vascular plants survive. Tree line in tropical regions is a consequence of low maximum temperatures throughout the year. However, the microclimate near the ground is warmer, allowing prostrate shrubs to grow at altitudes well above the highest trees.

Biota

Flora

Fir and hemlock: Rocky Mountain fir (Abies lasiocarpa, centre) among mountain hemlock (Tsuga mertensiana).

Pink mountain heather: Pink mountain heather (Phyllodoce species) growing.

Mountains in north temperate regions, such as those of North America, Europe, and northern Asia, generally have conifer-dominated forest on their lower slopes that gives way to alpine vegetation above. Typical conifers in these mountain regions are pines (Pinus), firs (Abies), spruces (Picea), and the deciduous larches (Larix). Some areas have broad-leaved deciduous trees, and a variety of smaller plants are found beneath the trees, especially in moister spots. For example, mountains in the northern half of Japan that are higher than 1,400 to 1,500 metres have a sub-arctic coniferous forest belt, the dominant trees all being conifers in the genera Abies, Picea, and Larix. Heathers, poppies, and the large carrot relative Oplopanax are a few of the other plants that grow in these forests. In some areas moorland vegetation is found, dominated by the moss Sphagnum. Birch (Betula) fringes the forest at its upper limit and occupies areas with a history of burning. In the Pacific Northwest of North America Pinus, Picea, and Abies usually dominate tree line forests. Aspens (Populus tremuloides) occur in places, especially those areas with a history of disturbance. Alders (Alnus) are found in avalanche tracks, and willows (Salix) are important species in wet places. Lupins (Lupinus), pasqueflowers (Anemone), and a large variety of daisies and low shrubs in the heather family are examples of the rich flora of smaller plants that grow beneath the trees and in meadows near the tree line.

Tree line forests in south temperate mountain regions also are dominated by only one or very few different types of tree at any site; the trees involved are usually broad-leaved rather than coniferous. For instance, most Australian mountains have tree line forests dominated by Eucalyptus, although a long history of widespread burning may be responsible to some extent for the prominence of this fire-tolerant tree.

In the tropics, by contrast, species-diverse forests that can be described as stunted evergreen rainforests typically grow as far as the uppermost limits of tree growth. This is the case in New Guinea, Southeast Asia, and East Africa; however, in parts of the tropical Andes, single species of Polylepis (of the rose family) often grow at altitudes above all other trees, especially on screes (rock debris that has accumulated at the base of a cliff).

Above the tree line, alpine vegetation comprises a variety of different subtypes including grasslands, mires, low heathlands, and crevice-occupying vegetation. For example, treeless alpine vegetation is found on mountains above 2,500 metres in central Japan, grading down to 1,400 metres in northern Hokkaido. Japanese stone pine (Pinus pumila), heathers, and grasses are particularly prominent. Like most other plants in this alpine vegetation, these plants have near relatives in the alpine areas of other mountainous, north temperate regions. The prostrate shrubs of the stone pine form dense, low thickets about one metre tall on ridges; they are mixed with deciduous shrubs of alder and service tree (Sorbus) in moister places. Other alpine communities occupy wet sites, where tall grassy meadows or bog communities often boast abundant tiny primroses (Primula nipponica). Stunted dwarf shrubs, especially members of the heather family and their relatives Arcterica, Vaccinium, Diapensia, and Empetrum occur where winter snow is blown from exposed surfaces. Conversely, in places where snow accumulates as deep drifts in sheltered spots and where it remains until late spring or summer, snowbed communities occur that are dominated by the heather Phyllodoce or by sedges (species of Carex), with many other small plants also present. Alpine deserts are also widespread in the high mountains of Japan, in places with marked soil instability associated with the effects of recent volcanic activity. While the plants surviving in such places are varied, some, like the violet Viola crassa, are typical of these harsh habitats.

Remarkably, the flora in the diverse array of alpine vegetation subtypes, such as in the above example, typically consists of a similar number of different plant species—about 200—in many regions both temperate and tropical. Furthermore, despite wide ecological and geographic contrasts, many of the same types of plant are found in most alpine regions. They are usually represented by different though related species in each region and on each mountain within regions. Gentians (Gentiana), plantains (Plantago), buttercups (Ranunculus), and members of the heather, grass, and sedge families are widespread examples.

However, some regional peculiarities exist both in alpine flora and in vegetation structure. One striking example concerns the large stem rosette plants found on several high tropical, but not temperate, mountains. These are giant herbs that reach three metres in height or beyond; they have persistent dead leaf bases that insulate the water-containing tissues of the stem from freezing conditions that can occur virtually every night in their very high (up to 4,300 metres), dry environments. Similar but unrelated stem rosette plants are found in the northern Andes (Espeletia and Puya) and on mountains in East Africa (Dendrosenecio and Lobelia), with other examples in Hawaii, Java, and the Himalayas. This emergence of the same characteristic among different species that are under the same environmental pressures on different continents is an example of convergent evolution.

Vegetation profile of tropical mountain lands.

Fauna

Mountain fauna is less distinctive than the flora of the same places and usually reflects the regional fauna. For example, the large mammals of North American mountain lands include deer, bears, wolves, and several large cats, all of which inhabit, or did before human invasion, the surrounding areas beyond the mountains. Some birds are tied to mountain habitats, such as the condors of the high ranges of California and the Andes. On certain mountains, flightless insects such as grasshoppers are a feature of interest, a phenomenon that is particularly pronounced on East African peaks such as Kilimanjaro.

As a result of their range of diverse topographic and climatic environments, and because evolution of cold-adapted biota has often proceeded independently on separate mountains in the same area, mountain regions are often noted as being centres of high biodiversity. The Caucasus Mountains in Asia provide one well-known example, while, in the tropics, the mountains of New Guinea contribute greatly to an enormous diversity of organisms, including some 20,000 plant species that represent 10 percent of the world's flora.

Population, Community Development and Structure

Population and community processes in temperate mountain regions, as in the rather similar Arctic environments, are influenced by the highly seasonal climate. As the winter snowpack melts, plants undergo a surge of growth and flowering, particularly in the alpine zone where the entire growing season is completed within about three months. Substantial food reserves in subterranean organs are used to generate mature, fertile shoots very rapidly, with growth in some cases beginning under the snow before melt is complete. Because almost the entire alpine flora blooms within the same period of only a few weeks and because alpine plants tend to have relatively large flowers, the floral display to be seen in temperate mountains in summer is often spectacular. In tropical mountains there is no such period of spectacular development that alternates with a longer season of enforced dormancy, and plants grow throughout the year unless their development is stopped by the onset of a dry season.

Animal activity similarly varies seasonally between regions. In temperate mountains there is a long period during which most birds and larger mammals migrate to lower altitudes. Some remaining mammals, such as the gophers of North American mountains, take advantage of the insulated environment beneath the snow where they make burrows and feed on subterranean plant organs.

In tropical mountains seasonal changes are much less pronounced, and this is reflected in animal reproduction. For example, birds on high mountains in New Guinea may breed throughout the year; however, because there is no seasonal flush of plant and insect growth creating a temporarily abundant source of food, they lay few eggs. Clutches of only one or two are normal there, by contrast with the five to eight eggs typically laid by many temperate mountain birds during their brief breeding season.

Biological Productivity

As stressful habitats for plants, mountain lands are not very productive environments. The biomass (dry weight of organic matter in an area) of the alpine vegetation on high temperate mountains, however, may be greater than it first appears because more than 10 times the amount of visible, aboveground biomass is present below the ground in the form of roots, rhizomes, tubers, and bulbs. By contrast, plants of the tropical alpine flora do not need to store food below ground, and less than half of the total biomass is located there.

Agricultural exploitation of mountain lands, therefore, is not very productive and generally is not intensive, being mainly confined to light or seasonal grazing by cattle, goats, and sheep. Where it occurs at moderate intensity, grazing can be very destructive to alpine vegetation, which cannot easily cope with disturbance in its already environmentally stressful state. Similarly, the physical disturbance associated with other human uses of high mountains, such as skiing and other forms of recreation, can be permanently damaging. Another concern is that atmospheric pollutants tend to become concentrated in snowfall. In temperate regions a pulse of polluting substances enters the alpine system with the annual snowmelt, bringing possibly detrimental consequences in this low-nutrient environment.

Aquatic Ecosystem

Aquatic ecosystems are any water-based environment in which plants and animals interact with the chemical and physical features of the aquatic environment. Aquatic ecosystems are generally divided into two types -the marine ecosystem and the freshwater ecosystem. The largest water ecosystem is the marine ecosystem, covering over 70 percent of the earth's surface. Oceans, estuaries, coral reefs and coastal ecosystems are the various kinds of marine ecosystems. Freshwater ecosystems cover less than 1 percent of the earth and are subdivided into lotic, lentic and wetlands.

Ocean Ecosystems

The earth has five major oceans: Pacific Ocean, Indian Ocean, Arctic Ocean, Atlantic Ocean and Southern (Antarctic) Ocean. Although the oceans are connected, each of them has unique species and features. The Pacific is the largest and deepest ocean and the Atlantic is second in size.

Oceans are home to different species of life. The waters of the Arctic and Southern Oceans are very cold, yet filled with life. The largest population of krill (small, shrimp-like marine creatures) lies under the ice of the Southern Ocean.

Estuaries

Estuaries are places where rivers meet the sea and may be defined as areas where salt water is diluted with fresh water. River mouths, coastal bays, tidal marshes and water bodies behind barrier beaches are some examples of estuaries. They are biologically productive as they have a special kind of water circulation that traps plant nutrients and stimulates primary production.

Coral Reefs

Coral reefs are the world's second richest ecosystems and have a wide diversity of plants and animals. As a result, coral reefs often are referred to as the rain forest of the oceans.

Coastal Systems

Land and water join to create the coastal ecosystems. These ecosystems have a distinct structure, diversity, and flow of energy. Plants and algae are found at the bottom of the coastal ecosystem. The fauna is diverse and consists of insects, snails, fish, crabs, shrimp, lobsters etc.

Lotic Ecosystems

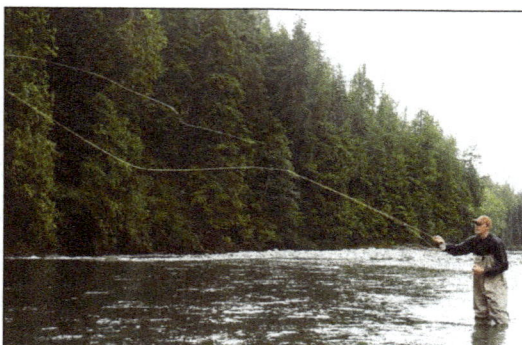

Lotic ecosystems are the systems with rapid flowing waters that move in a unidirectional way such as rivers and streams. These environments harbor numerous species of insects such as mayflies, stoneflies and beetles which have developed adapted features such as weighted cases to survive the environment. Several species of fishes such as eel, trout and minnow are found here. Various mammals such as beavers, otters and river dolphins inhabit lotic ecosystems.

Lentic Ecosystems

Lentic ecosystems include all standing water habitats such as lakes and ponds. These ecosystems are home to algae, rooted and floating-leaved plants and invertebrates such as crabs and shrimps. Amphibians such as frogs and salamanders and reptiles like alligators and water snakes are also found here.

Swamps and Wetlands

Wetlands are marshy areas and are sometimes covered in water which have a wide diversity of plants and animals. Swamps, marshes, and bogs are some examples in this regard. Plants such as black spruce and water lilies are commonly found in wetlands. The fauna consists of dragonflies and damselflies, birds such as Green Heron and fishes such as Northern Pike.

Freshwater Ecosystem

Freshwater ecosystems are a subset of Earth's aquatic ecosystems. They include lakes and ponds, rivers, streams, springs, bogs, and wetlands. They can be contrasted with marine ecosystems, which have a larger salt content. Freshwater habitats can be classified by different factors, including temperature, light penetration, nutrients, and vegetation.

Freshwater ecosystems can be divided into lentic ecosystems (still water) and lotic ecosystems (flowing water).

Limnology (and its branch freshwater biology) is a study about freshwater ecosystems. It is a part of hydrobiology.

Original attempts to understand and monitor freshwater ecosystems were spurred on by threats to human health (ex. Cholera outbreaks due to sewage contamination). Early monitoring focused on chemical indicators, then bacteria, and finally algae, fungi and protozoa. A new type of monitoring involves quantifying differing groups of organisms (macroinvertebrates, macrophytes and fish) and measuring the stream conditions associated with them.

Threats to Freshwater Ecosystems

Five broad threats to freshwater biodiversity include overexploitation, water pollution, flow modification, destruction or degradation of habitat, and invasion by exotic species. Recent extinction trends can be attributed largely to sedimentation, stream fragmentation, chemical and organic pollutants, dams, and invasive species. Common chemical stresses on freshwater ecosystem health include acidification, eutrophication and copper and pesticide contamination.

Extinction of Freshwater Fauna

Over 123 freshwater fauna species have gone extinct in North America since 1900. Of North American freshwater species, an estimated 48.5% of mussels, 22.8% of gastropods, 32.7% of crayfishes, 25.9% of amphibians, and 21.2% of fish are either endangered or threatened. Extinction rates of many species may increase severely into the next century because of invasive species, loss of keystone species, and species which are already functionally extinct (e.g., species which are not reproducing). Even using conservative estimates, freshwater fish extinction rates in North America are 877 times higher than background extinction rates (1 in 3,000,000 years). Projected extinction rates for freshwater animals are around five times greater than for land animals, and are comparable to the rates for rainforest communities.

Biomonitoring

Current freshwater biomonitoring techniques focus primarily on community structure, but some programs measure functional indicators like biochemical (or biological) oxygen demand, sediment oxygen demand, and dissolved oxygen. Macroinvertebrate community structure is commonly monitored because of the diverse taxonomy, ease of collection, sensitivity to a range of stressors, and overall value to the ecosystem. Additionally, algal community structure (often using diatoms) is measured in biomonitoring programs. Algae are also taxonomically diverse, easily collected, sensitive to a range of stressors, and overall valuable to the ecosystem. Algae grow very quickly and communities may represent fast changes in environmental conditions.

In addition to community structure, responses to freshwater stressors are investigated by experimental studies that measure organism behavioural changes, altered rates of growth, reproduction or mortality. Experimental results on single species under controlled conditions may not always reflect natural conditions and multi-species communities.

The use of reference sites is common when defining the idealized "health" of a freshwater ecosystem. Reference sites can be selected spatially by choosing sites with minimal impacts from human disturbance and influence. However, reference conditions may also be established temporally by using preserved indicators such as diatom valves, macrophyte pollen, insect chitin and fish scales can be used to determine conditions prior to large scale human disturbance. These temporal reference conditions are often easier to reconstruct in standing water than moving water because stable sediments can better preserve biological indicator materials.

Marine Ecosystem

Marine ecosystem is complex of living organisms in the ocean environment. Marine waters cover two-thirds of the surface of the Earth. In some places the ocean is deeper than Mount Everest is high; for example, the Mariana Trench and the Tonga Trench in the western part of the Pacific Ocean reach depths in excess of 10,000 metres (32,800 feet). Within this ocean habitat live a wide variety of organisms that have evolved in response to various features of their environs.

The Earth formed approximately 4.5 billion years ago. As it cooled, water in the atmosphere condensed and the Earth was pummeled with torrential rains, which filled its great basins, forming seas. The primeval atmosphere and waters harboured the inorganic components hydrogen, methane, ammonia, and water. These substances are thought to have combined to form the first organic compounds when sparked by electrical discharges of lightning. Some of the earliest known organisms are cyanobacteria (formerly referred to as blue-green algae). Evidence of these early photosynthetic prokaryotes has been found in Australia in Precambrian marine sediments called stromatolites that are approximately 3 billion years old. Although the diversity of life-forms observed in modern oceans did not appear until much later, during the Precambrian (about 4.6 billion to 542 million years ago) many kinds of bacteria, algae, protozoa, and primitive metazoa evolved to exploit the early marine habitats of the world. During the Cambrian Period (about 542 million to 488 million years ago) a major radiation of life occurred in the oceans. Fossils of familiar organisms such as cnidaria (e.g., jellyfish), echinoderms (e.g., feather stars), precursors of the fishes (e.g., the protochordate *Pikaia* from the Burgess Shale of Canada), and other vertebrates are found in marine sediments of this age. The first fossil fishes are found in sediments from the Ordovician Period (about 488 million to 444 million years ago). Changes in the physical conditions of the ocean that are thought to have occurred in the Precambrian—an increase in the concentration of oxygen in seawater and a buildup of the ozone layer that reduced dangerous ultraviolet radiation—may have facilitated the increase and dispersal of living things.

The Marine Environment

Geography, Oceanography and Topography

The shape of the oceans and seas of the world has changed significantly throughout the past 600 million years. According to the theory of plate tectonics, the crust of the Earth is made up of many dynamic plates. There are two types of plates—oceanic and continental—which float on the surface of the Earth's mantle, diverging, converging, or sliding against one another. When two plates diverge, magma from the mantle wells up and cools, forming new crust; when convergence occurs, one plate descends—i.e., is subducted—below the other and crust is resorbed into the mantle. Examples of both processes are observed in the marine environment. Oceanic crust is created along oceanic ridges or rift areas, which are vast undersea mountain ranges such as the Mid-Atlantic Ridge.

The shape of the ocean also is altered as sea levels change. During ice ages a higher proportion of the waters of the Earth is bound in the polar ice caps, resulting in a relatively low sea level. When the polar ice caps melt during interglacial periods, the sea level rises. These changes in sea level cause great changes in the distribution of marine environments such as coral reefs. For example, during the last Pleistocene Ice Age the Great Barrier Reef did not exist as it does today; the continental shelf on which the reef now is found was above the high-tide mark.

Marine organisms are not distributed evenly throughout the oceans. Variations in characteristics of the marine environment create different habitats and influence what types of organisms will inhabit them. The availability of light, water depth, proximity to land, and topographic complexity all affect marine habitats.

The availability of light affects which organisms can inhabit a certain area of a marine ecosystem. The greater the depth of the water, the less light can penetrate until below a certain depth there is no light whatsoever. This area of inky darkness, which occupies the great bulk of the ocean, is called the aphotic zone. The illuminated region above it is called the photic zone, within which are distinguished the euphotic and disphotic zones. The euphotic zone is the layer closer to the surface that receives enough light for photosynthesis to occur. Beneath lies the disphotic zone, which is illuminated but so poorly that rates of respiration exceed those of photosynthesis. The actual depth of these zones depends on local conditions of cloud cover, water turbidity, and ocean surface. In general, the euphotic zone can extend to depths of 80 to 100 metres and the disphotic zone to depths of 80 to 700 metres. Marine organisms are particularly abundant in the photic zone, especially the euphotic portion; however, many organisms inhabit the aphotic zone and migrate vertically to the photic zone every night. Other organisms, such as the tripod fish and some species of sea cucumbers and brittle stars, remain in darkness all their lives.

Marine environments can be characterized broadly as a water, or pelagic, environment and a bottom, or benthic, environment. Within the pelagic environment the waters are divided into the neritic province, which includes the water above the continental shelf, and the oceanic province, which includes all the open waters beyond the continental shelf. The high nutrient levels of the neritic province—resulting from dissolved materials in riverine runoff—distinguish this province from the oceanic. The upper portion of both the neritic and oceanic waters—the epipelagic zone—is where photosynthesis occurs; it is roughly equivalent to the photic zone. Below this zone lie the mesopelagic, ranging between 200 and 1,000 metres, the bathypelagic, from 1,000 to 4,000 metres, and the abyssalpelagic, which encompasses the deepest parts of the oceans from 4,000 metres to the recesses of the deep-sea trenches.

The benthic environment also is divided into different zones. The supralittoral is above the high-tide mark and is usually not under water. The intertidal, or littoral, zone ranges from the high-tide mark (the maximum elevation of the tide) to the shallow, offshore waters. The sublittoral is the environment beyond the low-tide mark and is often used to refer to substrata of the continental shelf, which reaches depths of between 150 and 300 metres. Sediments of the continental shelf that influence marine organisms generally originate from the land, particularly in the form of riverine runoff, and include clay, silt, and sand. Beyond the continental shelf is the bathyal zone, which occurs at depths of 150 to 4,000 metres and includes the descending continental slope and rise. The abyssal zone (between 4,000 and 6,000 metres) represents a substantial portion of the oceans. The deepest region of the oceans (greater than 6,000 metres) is the hadal zone of the deep-sea trenches. Sediments of the deep sea primarily originate from a rain of dead marine organisms and their wastes.

Physical and Chemical Properties of Seawater

The physical and chemical properties of seawater vary according to latitude, depth, nearness to land, and input of fresh water. Approximately 3.5 percent of seawater is composed of dissolved compounds, while the other 96.5 percent is pure water. The chemical composition of seawater reflects such processes as erosion of rock and sediments, volcanic activity, gas exchange with the atmosphere, the metabolic and breakdown products of organisms, and rain. In addition to carbon, the nutrients essential for living organisms include nitrogen and phosphorus, which are minor constituents of seawater and thus are often limiting factors in organic cycles of the ocean. Concentrations of phosphorus and nitrogen are generally low in the photic zone because they are rapidly taken up by marine organisms. The highest concentrations of these nutrients generally are found below 500 metres, a result of the decay of organisms. Other important elements include silicon and calcium (essential in the skeletons of many organisms such as fish and corals).

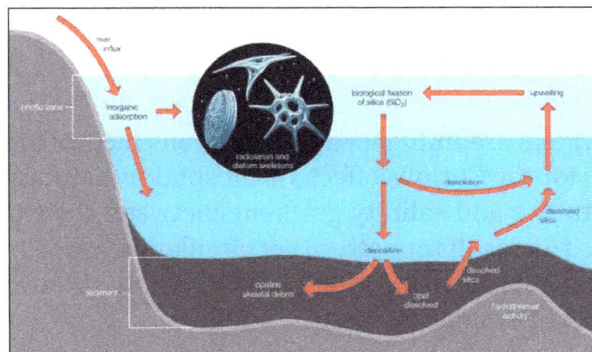

Cycling of silica in the marine environment.

Silicon commonly occurs in nature as silicon dioxide (SiO_2), also called silica. It cycles through the marine environment, entering primarily through riverine runoff. Silica is removed from the ocean by organisms such as diatoms and radiolarians that use an amorphous form of silica in their cell walls. After they die, their skeletons settle through the water column and the silica redissolves. A small number reach the ocean floor, where they either remain, forming a silaceous ooze, or dissolve and are returned to the photic zone by upwelling.

The chemical composition of the atmosphere also affects that of the ocean. For example, carbon dioxide is absorbed by the ocean and oxygen is released to the atmosphere through the activities of marine plants. The dumping of pollutants into the sea also can affect the chemical makeup of the ocean, contrary to earlier assumptions that, for example, toxins could be safely disposed of there.

The physical and chemical properties of seawater have a great effect on organisms, varying especially with the size of the creature. As an example, seawater is viscous to very small animals (less than 1 millimetre [0.039 inch] long) such as ciliates but not to large marine creatures such as tuna.

Marine organisms have evolved a wide variety of unique physiological and morphological features that allow them to live in the sea. Notothenid fishes in Antarctica are able to inhabit waters as cold as −2 °C (28 °F) because of proteins in their blood that act as antifreeze. Many organisms are able to achieve neutral buoyancy by secreting gas into internal chambers, as cephalopods do, or into swim bladders, as some fish do; other organisms use lipids, which are less dense than water, to achieve this effect. Some animals, especially those in the aphotic zone, generate light to attract prey. Animals in the disphotic zone such as hatchet fish produce light by means of organs called photophores to break up the silhouette of their bodies and avoid visual detection by predators. Many marine animals can detect vibrations or sound in the water over great distances by means of specialized organs. Certain fishes have lateral line systems, which they use to detect prey, and whales have a sound-producing organ called a melon with which they communicate. Tolerance to differences in salinity varies greatly: stenohaline organisms have a low tolerance to salinity changes, whereas euryhaline organisms, which are found in areas where river and sea meet (estuaries), are very tolerant of large changes in salinity. Euryhaline organisms are also very tolerant of changes in temperature. Animals that migrate between fresh water and salt water, such as salmon or eels, are capable of controlling their osmotic environment by active pumping or the retention of salts. Body architecture varies greatly in marine waters. The body shape of the cnidarian by-the-wind-sailor (Velella velella)—an animal that lives on the surface of the water (pleuston) and sails with the assistance of a modified flotation chamber—contrasts sharply with the sleek, elongated shape of the barracuda.

Ocean Currents

The movements of ocean waters are influenced by numerous factors including the rotation of the Earth (which is responsible for the Coriolis effect.) atmospheric circulation patterns that influence surface waters, and temperature and salinity gradients between the tropics and the polar region (thermohaline circulation). The resultant patterns of circulation range from those that cover great areas, such as the North Subtropical Gyre, which follows a path thousands of kilometres long, to small-scale turbulences of less than one metre.

Marine organisms of all sizes are influenced by these patterns, which can determine the range of a species. For example, krill (Euphausia superba) are restricted to the Antarctic Circumpolar Current. Distribution patterns of both large and small pelagic organisms are affected as well. Mainstream currents such as the Gulf Stream and East Australian Current transport larvae great distances. As a result cold temperate coral reefs receive a tropical infusion when fish and invertebrate larvae from the tropics are relocated to high latitudes by these currents. The successful recruitment of eels to Europe depends on the strength of the Gulf Stream to transport them from spawning sites in the Caribbean. Areas where the ocean is affected by nearshore features, such as estuaries, or areas in which there is a vertical salinity gradient (halocline) often exhibit intense biological activity. In these environments, small organisms can become concentrated, providing a rich supply of food for other animals.

Marine Biota

Marine biota can be classified broadly into those organisms living in either the pelagic environment (plankton and nekton) or the benthic environment (benthos). Some organisms, however, are benthic in one stage of life and pelagic in another. Producers that synthesize organic molecules exist in both environments. Single-celled or multicelled plankton with photosynthetic pigments are the producers of the photic zone in the pelagic environment. Typical benthic producers are microalgae (e.g., diatoms), macroalgae (e.g., the kelp Macrocystis pyrifera), or sea grass (e.g., Zostera).

Plankton

Plankton are the numerous, primarily microscopic inhabitants of the pelagic environment. They are critical components of food chains in all marine environments because they provide nutrition for the nekton (e.g., crustaceans, fish, and squid) and benthos (e.g., sea squirts and sponges). They also exert a global effect on the biosphere because the balance of components of the Earth's atmosphere depends to a great extent on the photosynthetic activities of some plankton.

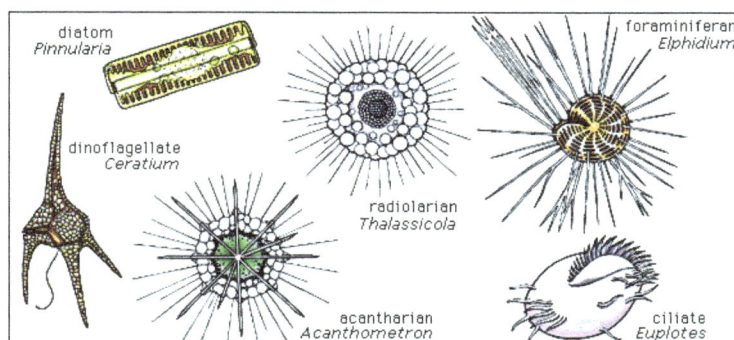

Representative plankton.

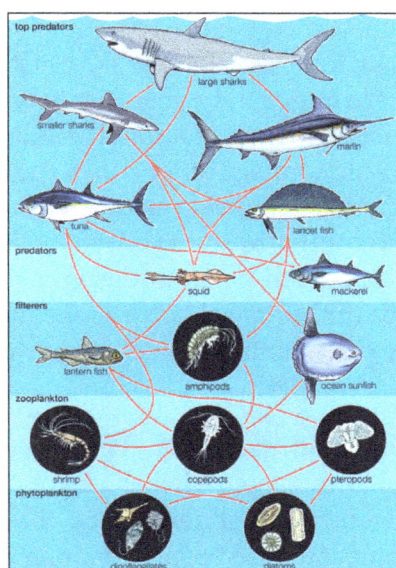

Generalized aquatic food web. Parasites, among the most diverse species in the food web, are not shown.

The term plankton means wandering or drifting, an apt description of the way most plankton spend their existence, floating with the ocean's currents. Not all plankton, however, are unable to control their movements, and many forms depend on self-directed motions for their survival.

Plankton range in size from tiny microbes (1 micrometre [0.000039 inch] or less) to jellyfish whose gelatinous bell can reach up to 2 metres in width and whose tentacles can extend over 15 metres. However, most planktonic organisms, called plankters, are less than 1 millimetre (0.039 inch) long. These microbes thrive on nutrients in seawater and are often photosynthetic. The plankton include a wide variety of organisms such as algae, bacteria, protozoans, the larvae of some animals, and crustaceans. A large proportion of the plankton are protists—i.e., eukaryotic, predominantly single-celled organisms. Plankton can be broadly divided into phytoplankton, which are plants or plantlike protists; zooplankton, which are animals or animal-like protists; and microbes such as bacteria. Phytoplankton carry out photosynthesis and are the producers of the marine community; zooplankton are the heterotrophic consumers.

Diatoms and dinoflagellates (approximate range between 15 and 1,000 micrometres in length) are two highly diverse groups of photosynthetic protists that are important components of the plankton. Diatoms are the most abundant phytoplankton. While many dinoflagellates carry out photosynthesis, some also consume bacteria or algae. Other important groups of protists include flagellates, foraminiferans, radiolarians, acantharians, and ciliates. Many of these protists are important consumers and a food source for zooplankton.

Zooplankton, which are greater than 0.05 millimetre in size, are divided into two general categories: meroplankton, which spend only a part of their life cycle—usually the larval or juvenile stage—as plankton, and holoplankton, which exist as plankton all their lives. Many larval meroplankton in coastal, oceanic, and even freshwater environments (including sea urchins, intertidal snails, and crabs, lobsters, and fish) bear little or no resemblance to their adult forms. These larvae may exhibit features unique to the larval stage, such as the spectacular spiny armour on the larvae of certain crustaceans (e.g., Squilla), probably used to ward off predators.

Important holoplanktonic animals include such lobsterlike crustaceans as the copepods, cladocerans, and euphausids (krill), which are important components of the marine environment because they serve as food sources for fish and marine mammals. Gelatinous forms such as larvaceans, salps, and siphonophores graze on phytoplankton or other zooplankton. Some omnivorous zooplankton such as euphausids and some copepods consume both phytoplankton and zooplankton; their feeding behaviour changes according to the availability and type of prey. The grazing and predatory activity of some zooplankton can be so intense that measurable reductions in phytoplankton or zooplankton abundance (or biomass) occur. For example, when jellyfish occur in high concentration in enclosed seas, they may consume such large numbers of fish larvae as to greatly reduce fish populations.

The jellylike plankton are numerous and predatory. They secure their prey with stinging cells (nematocysts) or sticky cells (colloblasts of comb jellies). Large numbers of the Portuguese man-of-war (Physalia), with its conspicuous gas bladder, the by-the-wind-sailor (Velella velella), and the small blue disk-shaped Porpita porpita are propelled along the surface by the wind, and after strong onshore winds they may be found strewn on the beach. Beneath the surface, comb jellies often abound, as do siphonophores, salps, and scyphomedusae.

The pelagic environment was once thought to present few distinct habitats, in contrast to the array of niches within the benthic environment. Because of its apparent uniformity, the pelagic realm was understood to be distinguished simply by plankton of different sizes. Small-scale variations in the pelagic environment, however, have been discovered that affect biotic distributions. Living and dead matter form organic aggregates called marine snow to which members of the plankton community may adhere, producing patchiness in biotic distributions. Marine snow includes structures such as aggregates of cells and mucus as well as drifting macroalgae and other flotsam that range in size from 0.5 millimetre to 1 centimetre (although these aggregates can be as small as 0.05 millimetre and as large as 100 centimetres). Many types of microbes, phytoplankton, and zooplankton stick to marine snow, and some grazing copepods and predators will feed from the surface of these structures. Marine snow is extremely abundant at times, particularly after plankton blooms. Significant quantities of organic material from upper layers of the ocean may sink to the ocean floor as marine snow, providing an important source of food for bottom dwellers. Other structures that plankton respond to in the marine environment include aggregates of phytoplankton cells that form large rafts in tropical and temperate waters of the world (e.g., cells of Oscillatoria [Trichodesmium] erthraeus) and various types of seaweed (e.g., Sargassum, Phyllospora, Macrocystis) that detach from the seafloor and drift.

Nekton

Nekton are the active swimmers of the oceans and are often the best-known organisms of marine waters. Nekton are the top predators in most marine food chains. The distinction between nekton and plankton is not always sharp. As mentioned above, many large marine animals, such as marlin and tuna, spend the larval stage of their lives as plankton and their adult stage as large and active members of the nekton. Other organisms such as krill are referred to as both micronekton and macrozooplankton.

The vast majority of nekton are vertebrates (e.g., fishes, reptiles, and mammals), mollusks, and crustaceans. The most numerous group of nekton are the fishes, with approximately 16,000 species. Nekton are found at all depths and latitudes of marine waters. Whales, penguins, seals, and icefish abound in polar waters. Lantern fish (family Myctophidae) are common in the aphotic zone along with gulpers (Saccopharynx), whalefish (family Cetomimidae), seven-gilled sharks, and others. Nekton diversity is greatest in tropical waters, where in particular there are large numbers of fish species.

The largest animals on the Earth, the blue whales (Balaenoptera musculus), which grow to 25 to 30 metres long, are members of the nekton. These huge mammals and other baleen whales (order Mysticeti), which are distinguished by fine filtering plates in their mouths, feed on plankton and micronekton as do whale sharks (Rhinocodon typus), the largest fish in the world (usually 12 to 14 metres long, with some reaching 17 metres). The largest carnivores that consume large prey include the toothed whales (order Odontoceti—for example, the killer whales, Orcinus orca), great white sharks (Carcharodon carcharias), tiger sharks (Galeocerdo cuvier), black marlin (Makaira indica), bluefin tuna (Thunnus thynnus), and giant groupers (Epinephelus lanceolatus).

Nekton form the basis of important fisheries around the world. Vast schools of small anchovies, herring, and sardines generally account for one-quarter to one-third of the annual harvest from the ocean. Squid are also economically valuable nekton. Halibut, sole, and cod are demersal (i.e., bottom-dwelling) fish that are commercially important as food for humans. They are generally caught in continental shelf waters. Because pelagic nekton often abound in areas of upwelling where the waters are nutrient-rich, these regions also are major fishing areas.

Benthos

Organisms are abundant in surface sediments of the continental shelf and in deeper waters, with a great diversity found in or on sediments. In shallow waters, beds of seagrass provide a rich habitat for polychaete worms, crustaceans (e.g., amphipods), and fishes. On the surface of and within intertidal sediments most animal activities are influenced strongly by the state of the tide. On many sediments in the photic zone, however, the only photosynthetic organisms are microscopic benthic diatoms.

Benthic organisms can be classified according to size. The macrobenthos are those organisms larger than 1 millimetre. Those that eat organic material in sediments are called deposit feeders (e.g., holothurians, echinoids, gastropods), those that feed on the plankton above are the suspension feeders (e.g., bivalves, ophiuroids, crinoids), and those that consume other fauna in the benthic assemblage are predators (e.g., starfish, gastropods). Organisms between 0.1 and 1 millimetre constitute the meiobenthos. These larger microbes, which include foraminiferans, turbellarians, and polychaetes, frequently dominate benthic food chains, filling the roles of nutrient recycler, decomposer, primary producer, and predator. The microbenthoses are those organisms smaller than 1 millimetre; they include diatoms, bacteria, and ciliates.

Organic matter is decomposed aerobically by bacteria near the surface of the sediment where oxygen is abundant. The consumption of oxygen at this level, however, deprives deeper layers of oxygen, and marine sediments below the surface layer are anaerobic. The thickness of the oxygenated layer varies according to grain size, which determines how permeable the sediment is to oxygen and the amount of organic matter it contains. As oxygen concentration diminishes, anaerobic processes come to dominate. The transition layer between oxygen-rich and oxygen-poor layers is called the redox discontinuity layer and appears as a gray layer above the black anaerobic layers. Organisms have evolved various ways of coping with the lack of oxygen. Some anaerobes release hydrogen sulfide, ammonia, and other toxic reduced ions through metabolic processes. The thiobiota, made up primarily of microorganisms, metabolize sulfur. Most organisms that live below the redox layer, however, have to create an aerobic environment for themselves. Burrowing animals generate a respiratory current along their burrow systems to oxygenate their dwelling places; the influx of oxygen must be constantly maintained because the surrounding anoxic layer quickly depletes the burrow of oxygen. Many bivalves (e.g., Mya arenaria) extend long siphons upward into oxygenated waters near the surface so that they can respire and feed while remaining sheltered from predation deep in the sediment. Many large mollusks use a muscular "foot" to dig with, and in some cases they use it to propel themselves away from predators such as starfish. The consequent "irrigation" of burrow systems can create oxygen and nutrient fluxes that stimulate the production of benthic producers (e.g., diatoms).

Not all benthic organisms live within the sediment; certain benthic assemblages live on a rocky substrate. Various phyla of algae—Rhodophyta (red), Chlorophyta (green), and Phaeophyta (brown)— are abundant and diverse in the photic zone on rocky substrata and are important producers. In intertidal regions algae are most abundant and largest near the low-tide mark. Ephemeral algae such as Ulva, Enteromorpha, and coralline algae cover a broad range of the intertidal. The mix of algae species found in any particular locale is dependent on latitude and also varies greatly according to wave exposure and the activity of grazers. For example, Ascophyllum spores cannot attach to rock in even a gentle ocean surge; as a result this plant is largely restricted to sheltered shores. The fastest-growing plant—adding as much as 1 metre per day to its length—is the giant kelp, Macrocystis pyrifera, which is found on subtidal rocky reefs. These plants, which may exceed 30 metres in length, characterize benthic habitats on many temperate reefs. Large laminarian and fucoid algae are also common on temperate rocky reefs, along with the encrusting (e.g., Lithothamnion) or short tufting forms (e.g., Pterocladia). Many algae on rocky reefs are harvested for food, fertilizer, and pharmaceuticals. Macroalgae are relatively rare on tropical reefs where corals abound, but Sargassum and a diverse assemblage of short filamentous and tufting algae are found, especially at the reef crest. Sessile and slow-moving invertebrates are common on reefs. In the intertidal and subtidal regions herbivorous gastropods and urchins abound and can have a great influence on the distribution of algae. Barnacles are common sessile animals in the intertidal. In the subtidal regions, sponges, ascidians, urchins, and anemones are particularly common where light levels drop and current speeds are high. Sessile assemblages of animals are often rich and diverse in caves and under boulders.

Reef-building coral polyps (Scleractinia) are organisms of the phylum Cnidaria that create a calcareous substrate upon which a diverse array of organisms live. Approximately 700 species of corals are found in the Pacific and Indian oceans and belong to genera such as Porites, Acropora, and Montipora. Some of the world's most complex ecosystems are found on coral reefs. Zooxanthellae are the photosynthetic, single-celled algae that live symbiotically within the tissue of corals and help to build the solid calcium carbonate matrix of the reef. Reef-building corals are found only in waters warmer than 18 °C; warm temperatures are necessary, along with high light intensity, for the coral-algae complex to secrete calcium carbonate. Many tropical islands are composed entirely of hundreds of metres of coral built atop volcanic rock.

Links between Pelagic Environments and the Benthos

Considering the pelagic and benthic environments in isolation from each other should be done cautiously because the two are interlinked in many ways. For example, pelagic plankton are an important source of food for animals on soft or rocky bottoms. Suspension feeders such as anemones and barnacles filter living and dead particles from the surrounding water while detritus feeders graze on the accumulation of particulate material raining from the water column above. The molts of crustaceans, plankton feces, dead plankton, and marine snow all contribute to this rain of fallout from the pelagic environment to the ocean bottom. This fallout can be so intense in certain weather patterns—such as the El Niño condition—that benthic animals on soft bottoms are smothered and die. There also is variation in the rate of fallout of the plankton according to seasonal cycles of production. This variation can create seasonality in the abiotic zone where there is little or no variation in temperature or light. Plankton form marine sediments, and many types of fossilized protistan plankton, such as foraminiferans and coccoliths, are used to determine the age and origin of rocks.

Organisms of the Deep-sea Vents

Producers were discovered in the aphotic zone when exploration of the deep sea by submarine became common in the 1970s. Deep-sea hydrothermal vents now are known to be relatively common in areas of tectonic activity (e.g., spreading ridges). The vents are a nonphotosynthetic source of organic carbon available to organisms. A diversity of deep-sea organisms including mussels, large bivalve clams, and vestimentiferan worms are supported by bacteria that oxidize sulfur (sulfide) and derive chemical energy from the reaction. These organisms are referred to as chemoautotrophic, or chemosynthetic, as opposed to photosynthetic, organisms. Many of the species in the vent fauna have developed symbiotic relationships with chemoautotrophic bacteria, and as a consequence the megafauna are principally responsible for the primary production in the vent assemblage. The situation is analogous to that found on coral reefs where individual coral polyps have symbiotic relationships with zooxanthellae. In addition to symbiotic bacteria there is a rich assemblage of free-living bacteria around vents. For example, *Beggiatoas*-like bacteria often form conspicuous weblike mats on any hard surface; these mats have been shown to have chemoautotrophic metabolism. Large numbers of brachyuran (e.g., *Bythograea*) and galatheid crabs, large sea anemones (e.g., *Actinostola callasi*), copepods, other plankton, and some fish—especially the eelpout *Thermarces cerberus*—is found in association with vents.

Hydrothermal mussels Galatheid crabs and shrimp grazing on the bacterial filaments that grow on the shells of the hydrothermal mussels covering the Northwest Eifuku volcano in the Mariana Arc region.

Structure of Marine Assemblages

Distribution and Dispersal

The distribution patterns of marine organisms are influenced by physical and biological processes in both ecological time (tens of years) and geologic time (hundreds to millions of years). The shapes of the Earth's oceans have been influenced by plate tectonics, and as a consequence the distributions of fossil and extant marine organisms also have been affected. Vicariance theory argues that plate tectonics has a major role in determining biogeographic patterns. For example, Australia was once—90 million years ago—close to the South Pole and had few coral reefs. Since then Australia has been moving a few millimetres each year closer to the Equator. As a result of this movement and local oceanographic conditions, coral reef environments are extending ever so slowly southward. Dispersal may also have an important role in biogeographic patterns of abundance. The importance of dispersal varies greatly with local oceanographic features, such as the direction and intensity of currents and the biology of the organisms. Humans can also have an impact on patterns of distribution and the extinction of marine organisms. For example, fishing intensity in the Irish Sea was based on catch limits set for cod with no regard for the biology of other species. One consequence of this practice was that the local skate, which had a slow reproductive rate, was quickly fished to extinction.

A characteristic of many marine organisms is a bipartite life cycle, which can affect the dispersal of an organism. Most animals found on soft and hard substrata, such as lobsters, crabs, barnacles, fish, polychaete worms, and sea urchins, spend their larval phase in the plankton and in this phase are dispersed most widely. The length of the larval phase, which can vary from a few minutes to hundreds of days, has a major influence on dispersal. For example, wrasses of the genus *Thalassoma* have a long larval life, compared with many other types of reef fish, and populations of these fish are well dispersed to the reefs of isolated volcanic islands around the Pacific. The bipartite life cycle of algae also affects their dispersal, which occurs through algal spores. Although in general, spores disperse only a short distance from adult plants, limited swimming abilities—*Macrocystis* spores have flagella—and storms can disperse spores over greater distances.

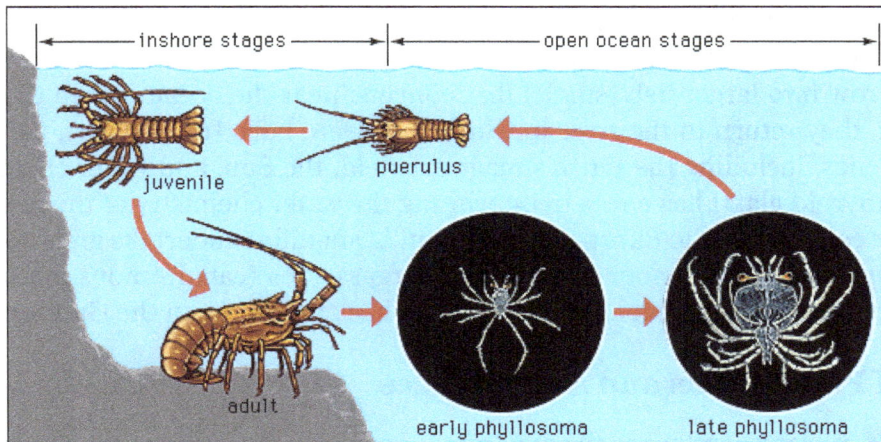

Life cycle of a palinurid lobster.

Migrations of Marine Organisms

The migrations of plankton and nekton throughout the water column in many parts of the world are well described. Diurnal vertical migrations are common. For example, some types of plankton, fish, and squid remain beneath the photic zone during the day, moving toward the surface after dusk and returning to the depths before dawn. It is generally argued that marine organisms migrate in response to light levels. This behaviour may be advantageous because by spending the daylight hours in the dim light or darkness beneath the photic zone plankton can avoid predators that locate their prey visually. After the Sun has set, plankton can rise to the surface waters where food is more abundant and where they can feed safely under the cover of darkness.

Larval forms can facilitate their horizontal transport along different currents by migrating vertically. This is possible because currents can differ in direction according to depth (e.g., above and below haloclines and thermoclines), as is the case in estuaries.

In coastal waters many larger invertebrates (e.g., mysids, amphipods, and polychaete worms) leave the cover of algae and sediments to migrate into the water column at night. It is thought that these animals disperse to different habitats or find mates by swimming when visual predators find it hard to see them. In some cases only one sex will emerge at night, and often that sex is morphologically better suited for swimming.

Horizontal migrations of fish that span distances of hundreds of metres to tens of kilometres are common and generally related to patterns of feeding or reproduction. Tropical coral trout (Plectropomus species) remain dispersed over a reef for most of the year, but adults will aggregate at certain locations at the time of spawning. Transoceanic migrations (greater than 1,000 kilometres) are observed in a number of marine vertebrates, and these movements often relate to require-ments of feeding and reproduction. Bluefin tuna (Thunnus thynnus) traverse the Atlantic Ocean in a single year; they spawn in the Caribbean then swim to high latitudes of the Atlantic to feed on the rich supply of fish. Turtles and sharks also migrate great distances.

Fish that spend their lives in both marine and freshwater systems (diadromous animals) exhibit some of the most spectacular migratory behaviour. Anadromous fishes (those that spend most of their lives in the sea but migrate to fresh water to spawn) such as Atlantic salmon (Salmo salar) also have unique migratory patterns. After spawning, the adults die. Newly hatched fish (alevin) emerge from spawned eggs and develop into young fry that move down rivers toward the sea. Ju-veniles (parr) grow into larger fish (smolt) that convene near the ocean. When the adult fish are ready to spawn, they return to the river in which they were born (natal river), using a variety of environmental cues, including the Earth's magnetic field, the Sun, and water chemistry. It is be-lieved that the thyroid gland has a role in imprinting the water chemistry of the natal river on the fish. Freshwater eels such as the European eel (Anguilla anguilla) undertake great migrations from fresh water to spawn in the marine waters of the Sargasso Sea (catadromous migrations), where they die. Eel larvae, called leptocephalus larvae, drift back to Europe in the Gulf Stream.

Dynamics of Populations and Assemblages

A wide variety of processes influence the dynamics of marine populations of individual species and the composition of assemblages (e.g., collections of populations of different species that live in the same area). With the exception of marine mammals such as whales, fish that bear live young (e.g., embiotocid fish), and brooders (i.e., fauna that incubate their offspring until they emerge as larvae or juveniles), most marine organisms produce a large number of offspring of which few survive. Processes that affect the plankton can have a great influence on the numbers of young that survive to be recruited, or relocated, into adult populations. The survival of larvae may depend on the abun-dance of food at various times and in various places, the number of predators, and oceanographic features that retain larvae near suitable nursery areas. The number of organisms recruited to ben-thic and pelagic systems may ultimately determine the size of adult populations and therefore the relative abundance of species in marine assemblages. However, many processes can affect the sur-vival of organisms after recruitment. Predators eat recruits, and mortality rates in prey species can vary with time and space, thus changing original population patterns established in recruitment.

Patterns of colonization and succession can have a significant impact on benthic assemblages. For example, when intertidal reefs are cleared experimentally, the assemblage of organisms that colonize the bare space often reflects the types of larvae available in local waters at the time. Tube worms may dominate if they establish themselves first; if they fail to do so, algal spores may col-onize the shore first and inhibit the settlement of these worms. Competition between organisms may also play a role. Long-term data gathered over periods of more than 25 years from coral reefs have demonstrated that some corals (e.g., Acropora cytherea) competitively overgrow neigh-bouring corals. Physical disturbance from hurricanes destroys many corals, and during regrowth competitively inferior species can coexist with normally dominant species on the reef. Chemical defenses of sessile organisms also can deter the growth or cause increased mortality of organisms that settle on them. Ascidian larvae (e.g., Podoclavella) often avoid settling on sponges (e.g., My-cale); when this does occur, the larvae rarely reach adulthood.

Although the processes that determine species assemblages may be understood, variations occur in the composition of the plankton that make it difficult to predict patterns of colonization with great accuracy.

Biological Productivity

Primary productivity is the rate at which energy is converted by photosynthetic and chemosynthetic autotrophs to organic substances. The total amount of productivity in a region or system is gross primary productivity. A certain amount of organic material is used to sustain the life of producers; what remains is net productivity. Net marine primary productivity is the amount of organic material available to support the consumers (herbivores and carnivores) of the sea. The standing crop is the total biomass (weight) of vegetation. Most primary productivity is carried out by pelagic phytoplankton, not benthic plants.

Most primary producers require nitrogen and phosphorus, which are available in the ocean as nitrate, nitrite, ammonia, and phosphorus. The abundances of these molecules and the intensity and quality of light exert a major influence on rates of production. The two principal categories of producers (autotrophs) in the sea are pelagic phytoplankton and benthic microalgae and macroalgae. Benthic plants grow only on the fringe of the world's oceans and are estimated to produce only 5 to 10 percent of the total marine plant material in a year. Chemoautotrophs are the producers of the deep-sea vents.

Primary productivity is usually determined by measuring the uptake of carbon dioxide or the output of oxygen. Production rates are usually expressed as grams of organic carbon per unit area per unit time. The productivity of the entire ocean is estimated to be approximately 16×10^{10} tons of carbon per year, which is about eight times that of the land.

Pelagic Food Chain

Food chains in coastal waters of the world are generally regulated by nutrient concentrations. These concentrations determine the abundance of phytoplankton, which in turn provide food for the primary consumers, such as protozoa and zooplankton that the higher-level consumers—fish, squid, and marine mammals—prey upon. It had been thought that phytoplankton in the 5- to 100-micrometre size range were responsible for most of the primary production in the sea and that grazers such as copepods controlled the numbers of phytoplankton. Data gathered since 1975, however, indicate that the system is much more complex than this. It is now thought that most primary production in marine waters of the world is accomplished by single-celled 0.5- to 10-micrometre phototrophs (bacteria and protists). Moreover, heterotrophic protists (phagotrophic protists) are now viewed as the dominant controllers of both bacteria and primary production in the sea. Current models of pelagic marine food chains picture complex interactions within a microbial food web. Larger metazoans are supported by the production of autotrophic and heterotrophic cells.

Upwelling

The most productive waters of the world are in regions of upwelling. Upwelling in coastal waters brings nutrients toward the surface. Phytoplankton reproduce rapidly in these conditions, and grazing zooplankton also multiply and provide abundant food supplies for nekton. Some of the world's richest fisheries are found in regions of upwelling—for example, the temperate waters off

Peru and California. If upwelling fails, the effects on animals that depend on it can be disastrous. Fisheries also suffer at these times, as evidenced by the collapse of the Peruvian anchovy industry in the 1970s. The intensity and location of upwelling are influenced by changes in atmospheric circulation, as exemplified by the influence of El Niño conditions.

Seasonal Cycles of Production

Cycles of plankton production vary at different latitudes because seasonal patterns of light and temperature vary dramatically with latitude. In the extreme conditions at the poles, plankton populations crash during the constant darkness of winter and bloom in summer with long hours of light and the retreat of the ice field. In tropical waters, variation in sunlight and temperature is slight, nutrients are present in low concentrations, and planktonic assemblages do not undergo large fluctuations in abundance. There are, however, rapid cycles of reproduction and high rates of grazing and predation that result in a rapid turnover of plankton and a low standing crop. In temperate regions plankton abundance peaks in spring as temperature and the length and intensity of daylight increase. Moreover, seasonal winter storms usually mix the water column, creating a more even distribution of the nutrients, which facilitates the growth of phytoplankton. Peak zooplankton production generally lags behind that of phytoplankton, while the consumption of phytoplankton by zooplankton and phagotrophic protists is thought to reduce phytoplankton abundance. Secondary peaks in abundance occur in autumn. Seasonal peaks of some plankton are very conspicuous, and the composition of the plankton varies considerably. In spring and early summer many fish and invertebrates spawn and release eggs and larvae into the plankton, and, as a result, the meroplanktonic component of the plankton is higher at these times. General patterns of plankton abundance may be further influenced by local conditions. Heavy rainfall in coastal regions (especially areas in which monsoons prevail) can result in nutrient-rich turbid plumes (i.e., estuarine or riverine plumes) that extend into waters of the continental shelf. Changes in production, therefore, may depend on the season, the proximity to fresh water, and the timing and location of upwelling, currents, and patterns of reproduction.

Forest Ecosystem

The forest ecosystem contains many smaller habitats.

A forest ecosystem is defined as an area dominated by trees and other woody plants. Forests aren't only trees, however. Healthy forests have a lot going on in them, and many different species of both animals and plants that call them home. There are many different types of forests in the world, ranging from tropical rain forests to the dense sub-polar taiga. To truly understand a forest ecosystem, it is easiest to break it down into the five layers that most healthy forests have. Animals that live in a forest move between the layers to feed and hunt.

Canopy

The canopy section of a forest is the very top, and consists of the tallest, oldest trees, which can reach heights of 150 to 200 feet. This layer is the harshest of the five layers because it is exposed to everything that nature throws its way. It gets whipped by the wind, exposed to the sun without shade, receives the brunt of downpours, and is the most likely to be struck by lightning. Animal that live in this layer are those adapted to living tough, and include birds, tree frogs, snakes, lizards and hard-bodied insects.

Understory

The understory is the layer just below the canopy, and consists of those trees that are still growing but haven't reached full height. This environment is protected from the elements somewhat by the canopy layer, and is therefore less harsh. Trees in the understory are growing slower because they have less light, and tend to be a bit thinner in foliage. There is a greater variety of animals that live in this layer, including birds, butterflies and caterpillars, frogs and tree mammals like squirrels and raccoons, in the north, and monkeys, in the tropics.

Shrub Layer

The shrub layer is the next level down, and is dominated by woody plants that never grow very tall. Some of these are very young trees or trees that remain shorter, but most are shrubs, which are woody plants that have more than one stem. Shrubs can get as tall as 15 to 20 feet, but most top out at around 10, and many are shorter than that. Lichens can grow on tree bark between the shrub layer and the understory, and animal life also thrives. The shrub layer is home to many different kinds of insects and spiders, birds, snakes and lizards.

Herbaceous Layer and Forest Floor

The herbaceous layer is the layer just above the forest floor, and consists of tree seedlings and non-woody plants. These include mosses and a variety of flowers. The forest floor consists of the leaf litter-- a thick bed of leaves dropped from the trees and the soil. These layers are the backbone of the forest. Without good soil, trees have nothing to root into, and in the north, the leaf litter acts as insulation for tree roots and soil-based animals. Hornets, butterflies, birds, worms, slugs, snails, centipedes, millipedes, spiders, snakes live at this level, as well as billions of microbes, all of which contribute to soil health.

Types

Temperate Forest Ecosystem

The temperate forest ecosystem is very important on Earth. Temperate forests are in regions where the climate changes a lot from summer to winter. Tropical rain forests are in regions where the climate stays constant all year long. Temperate forests are almost always made of two types of trees, deciduous and evergreen. Deciduous trees are trees that lose their leaves in the winter.

Evergreens are trees that keep them all year long, like pine trees. Forests can either be one or the other, or a combination of both. A fourth kind of forest is a temperate rain forest. These are found in California, Oregon and Washington in the United States.

These forests are made of redwoods and sequoias, the tallest trees in the world. The amount of rainfall in an area determines if a forest is present. If there is enough rain to support trees, then a forest will usually develop. Otherwise, the region will become grasslands.

The Tropical Rain Forest Ecosystem

Tropical rain forests are one of the most important areas on Earth. These special ecosystems are homes to thousands of species animals and plants. Contrary to popular belief, rain forests are not only densely packed plants, but are also full of tall trees that form a ceiling from the Sun above. This ceiling keeps smaller plants from growing. Areas where sunlight can reach the surface are full of interesting plants.

Do They are so named because they receive a lot of rain – an average of 80 inches a year. The temperature doesn't change very much during the year. It is always warm and muggy. The famous Amazon jungle is located in Brazil, in South America. This particular forest is called the Neotropics. Other large blocks are located in Central and West Africa.

Insects of the Tropical Rain Forest

The most feared and well known spider in the world resides in the jungle. Tarantulas are one of the creepiest animals you will ever see. Most species of tarantula have poisonous fangs for killing prey and for protection.

Although some are life-threatening to humans, others are harmless. Army ants are just one species of ant in the rain forest. They are called army ants because they march in a long, thick line through the jungle. They only stop when the young larvae reach pupil stage. Once the queen lays its eggs, the ants start marching again.

Beautiful butterflies fill the forest, but at one time these insects weren't so pretty. Butterflies start out as caterpillars, which tend to be a tad on the ugly side. They go through metamorphosis, which is the process of changing into a butterfly. Centipedes aren't so lucky. They don't turn into butterflies, but instead roam the forest looking for food. Some centipedes use poison to kill their prey.

Tropical Rain Forest Birds

The birds of the rain forest are the most beautiful in the world. A wide range of colors can be seen darting through the trees as the forest tops come to life. Many species of tropical birds are kept as pets because of their looks.

Hundreds of species of parrot live in the rain forest. The scarlet macaw is just one of these. It is also one of the longest, stretching to a length of 3 feet from its head to the tip of its tail. When these macaws eat a poisonous fruit, they eat a special type of clay that neutralizes the poison.

Toucans are also very interesting birds. They have large beaks that they use to reach fruit they can't get to. Scientists estimate there are 33 species of toucan in the rain forest. Not every tropical bird was blessed with looks. The hoatzin looks more like a peacock without the pretty tail.

Hoatzins are terrible flyers – crash landings are common practice. The brown kiwi is a flightless bird that looks more like a rodent with a long beak and feathers. Kiwis live on the ground instead of the trees. They have special claws used for running, digging and defence.

Tropical Rain Forest Mammals

Birds aren't the only creatures that fly through the rain forests. Several species of flying mammals live in the jungle. From the harmless fruit bat to the unique flying squirrel, the tropical rain forests are full of surprises.

The Indian flying fox is one of the largest bats in the world. Its wings can spread out to 5 feet in width. Unlike bats in other parts of the world, these bats do not live in caves. They prefer to .hang in trees during the day. Hundreds or even thousands of bats can be spotted in a single tree.

Vampire bats live in the Amazon jungle in South America. The famous stories of blood-sucking bats probably originated here. These bats do in fact drink the blood of their victims. They usually attack farm animals, but have also enjoyed the blood of humans. But vampire bats only drink a very small amount of fluid.

Tropical Rain Forest Reptiles

The tropical rain forests of the world are full of reptiles. Reptiles are cold blooded, which means their body temperature depends on their environment. So, it is important for them to stay in warm climates. Snakes are reptiles, and the rain forests are home to many. The mamba family is the most poisonous of all. They kill their prey by injecting poison with their sharp fangs.

Anacondas make up another snake family. They are some of the longest creatures in the world, as they can reach 30 feet in length. Anacondas prefer to wrap themselves around their prey and squeeze, rather than inject poison. Anacondas swallow their prey whole and sleep while the food is digesting. Chameleons are interesting lizards that can change color.

This enables them to blend in with their surroundings. Not only is this a great disguise from predators, it is also an easy way to sneak up on their prey. Chameleons only eat insects. Geckos are very neat creatures. The flying gecko can glide from tree to tree to escape from predators. Their grip is so strong, that if you tried to pull one off a window, the glass would break before the gecko would let go.

Tropical Rain Forest Primates

Monkeys and their cousins are all primates. Humans are also primates. There are many species of monkeys in the tropical rain forests of the world. Monkeys can be divided into two groups: new world monkeys and old world monkeys. New world monkeys live only in South and Central America. Spider monkeys live in the rain forests in the Andes Mountains.

They look very strange with their long noses. Spider monkeys eat mostly fruit and nuts, so they are called frugivores. They are joined by the howler monkeys. These primates are so named because they have a special sac that makes their sounds louder.

Old world monkeys live only in Africa and Asia. The colobus monkey is one such kind. These monkeys are called foliovores because they eat leaves. They live in small groups of 15, but other primates live in larger groups of up to 200. There are too many species. Chimpanzees, orangutans and gorillas are all called pongids. These primates are more famous than the others. Gorillas are too big to climb trees, so they are found on the forest floor.

Boreal or Taiga Forests

The boreal forest ecosystem is the contiguous green belt of conifer and deciduous trees that encircles a large portion of the Northern Hemisphere. In North America, the boreal forest stretches across most of northern Canada and into Alaska. It has long been identified as one of the world's great forest ecosystems.

This forest ecosystem covers roughly 35% of Canada's land mass and is the single largest land based ecosystem in North America. It also contains a significant proportion of Canada's biodiversity and has long been recognized as an important global carbon sink.

Although the boreal is relatively unknown, it is important as the "great lung" of North America, "breathing in" carbon dioxide and "exhaling" oxygen into the atmosphere. In short, the boreal forest manages to do what the rain forest of the Amazon does but with only the fraction of the flora and fauna.

This forest ecosystem houses the largest and smallest mammal species (wood bison & pygmy shrews) of the North American continent. The Boreal forest has many things: great lakes and northern rivers; vast bogs, fens and other organic wetlands. The rich wildlife diversity of the Boreal is a joy to behold: woodland caribou and lynx; whooping cranes and wood bison; northern owls; woodpeckers with three rather than four toes; colorful wood warblers.

The Boreal has more than 5,000 species of conspicuous and colorful fungi, celebrated far more in Scandinavia and Siberia than in North America. Then there are the precious old-growth forests, the richest and most biologically diverse of the Boreal forest communities that are essential for so many Boreal species.

Structure of Forest Ecosystems

Different organisms exist within the forest layers. These organisms interact with each other and their surroundings. Each organism has a role or niche in sustaining the ecosystem.

Some provide food for other organisms; others provide shelter or control populations through predation.

Producers

All living organisms' intake energy in order to survive- In a forest ecosystem, trees and other plants get their energy from sunlight. Plants produce their own food, in the form of carbohydrates. Plants are, therefore, called the primary producers, since they produce the basic foodstuffs for other organisms within food chains and food webs. Photosynthesis is the chemical reaction that allows plants to produce their own food.

Consumers

Animals cannot produce their own food. They must consume food sources for die energy they need to survive. All animals, including mammals, insects, and birds, are called consumers. Consumers rely on plants and other animals as a food source.

Primary consumers only eat plants and are referred to as herbivores. Secondary consumers are referred to as carnivores and feed on herbivores. Tertiary consumers are carnivores that feed on other carnivores. Omnivores eat both plant and animal matter.

Decomposers

Leaves, needles, and old branches fall to the forest floor as trees grow. Eventually all plants and animals die. These materials are decomposed by worms, microbes, fungi, ants, and other bugs.

Decomposers break these items down into their smallest primary elements to be used again. Decomposers are important in that they sustain the nutrient cycle of ecosystems.

Humans are Part of Forest Ecosystem

Humans are consumers. We get food and materials from forests. Because of this, we are a part of the forest ecosystem. Human consumption alters forest ecosystems. Human intervention may be necessary to sustain forest communities under the increased pressure of human use.

Components of Ecosystem

The structure of an ecosystem is basically a description of the species of organisms that are present, including information on their life histories, populations and distribution in space. It is a guide to who's who in the ecosystem. It also includes descriptive information on the non-living (physical) features of environment, including the amount and distribution of nutrients.

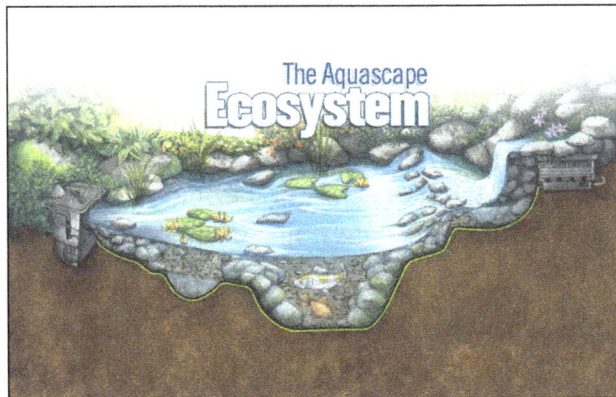

The structure of ecosystem provides information about the range of climatic conditions that prevail in the area. From structural point of view all ecosystems consist of following four basic components.

Abiotic Substances

These include basic inorganic and organic compounds of the environment or habitat of the organism. The inorganic components of an ecosystem are carbon dioxide, water, nitrogen, calcium, phosphate, all of which are involved in matter cycles (biogeochemical cycles).

The organic components of an ecosystem are proteins, carbohydrates, lipids and amino acids, all of which are synthesized by the biota (flora and fauna) of an ecosystem and are reached to ecosystem as their wastes, dead remains, etc. The climate, temperature, light, soil, etc. are other abiotic components of the ecosystem.

Producers

Producers are autotrophic organisms like chemosynthetic and photosynthetic bacteria, blue green algae, algae and all other green plants. They are called ecosystem producers because they capture energy from non-organic sources, especially light, and store some of the energy the form of chemical bonds, for the later use.

Algae of various types are the most important producers of aquatic ecosystems, although in estuaries and marshes, grasses may be important as producers. Terrestrial ecosystems have trees, shrubs, herbs, grasses, and mosses that contribute with varying importance to the production of the ecosystem.

Since heterotrophic organisms depend on plants and other autotrophic Organisms like bacteria and algae for their nutrition, the amount of energy that the producers capture, sets the limit on the availability of energy for the ecosystem. Thus, when a green plant captures a certain amount of energy from sunlight, it is said to "produce" the energy for the ecosystem.

Consumers

They are heterotrophic organisms in the ecosystem which eat other living creatures. There are herbivores, which eat plants, and carnivores, which eat other animals. They are also called phagotrophs or macroconsumers. Sometimes herbivores are called primary macroconsumers and carnivores are called secondary Macroconsumers.

Reducers or Decomposers

Reducers, decomposers, saprotrophs or Macro consumers are heterotrophic organisms that breakdown dead and waste matter. Fungi and certain bacteria are the prime representatives of this category. Enzymes are secreted by their cells into or onto dead plant and animal debris. These chemicals digest the dead organism into smaller bits or molecules, which can be absorbed by the fungi or bacteria (saprotrophs).

The decomposers take the energy and matter that they harvest during this feeding process for their own metabolism. Heat is liberated in each chemical conversion along the metabolic pathway.

No ecosystem could function long without decomposers. Dead organisms would pile up without rotting, as would waste products. It would not be long before an essential element, phosphorus, for example, would be first in short supply and then gone altogether, because the dead corpses littering the landscape would be hoarding the entire supply.

Thus, the importance of the decomposers to the ecosystem is that they tear apart organisms and in their metabolic processes release to the environment atoms and molecules that can be reused again by autotrophic organisms. They are not important to the ecosystem from the energy point of view but from the material (nutrient) point of view. Energy cannot be recycled, but matter can be.

Energy must be fed into ecosystem to keep up with the dissipation of heat or the increase in entropy.

References

- Ecosystem: nationalgeographic.org, Retrieved 7 March, 2019

- Desert-ecosystem: importantindia.com, Retrieved 11 January, 2019

- Desert-ecosystem: conserve-energy-future.com, Retrieved 21 August, 2019

- Desert-ecosystem: importantindia.com, Retrieved 10 April, 2019

- Mountain-ecosystem, science: britannica.com, Retrieved 14 March, 2019

- Types-aquatic-ecosystems: sciencing.com, Retrieved 4 May, 2019

- Stevenson, R. Jan; Smol, John P. (2003), "use of algae in environmental assessments", Freshwater Algae of North America, Elsevier, pp. 775–804, doi:10.1016/b978-012741550-5/50024-6, ISBN 9780127415505

- Marine-ecosystem, science: britannica.com, Retrieved 12 February, 2019

- Forest-ecosystem: classroom.synonym.com, Retrieved 20 June, 2019

- Forest-ecosystem-types-characteristic-features-and-structure, ecosystem, environment: yourarticlelibrary.com, Retrieved 23 April, 2019

- Basic-components-of-ecosystem, environment: yourarticlelibrary.com, Retrieved 3 July, 2019

Chapter 3

Energy Flow in Ecosystems

The flow of energy within a food chain is known as energy flow. The energy flows within the food web from the producers, also known as autotrophs to the consumers, who are also called heterotrophs. This chapter has been carefully written to provide an easy understanding of the varied types of food chains as well as ecological pyramids.

Energy flow is the one of the most fundamental processes and it is common to all the ecosystems. It is basically the movement of energy in an ecosystem through a series of organisms.

Ecosystem Energy

Organisms can be either producers or consumers in terms of energy flow through an ecosystem. Producers convert energy from the environment into carbon bonds, such as those found in the sugar glucose. Plants are the most obvious examples of producers; plants take energy from sunlight and use it to convert carbon dioxide into glucose (or other sugars). Algae and cyanobacteria are also photo- synthetic producers, like plants.

Other producers include bacteria living around deep-sea vents. These bacteria take energy from chemicals coming from the Earth's interior and use it to make sugars. Other bacteria living deep underground can also produce sugars from such inorganic sources. Another word for producers is autotrophs.

Routes of usage

Consumers get their energy from the carbon bonds made by the producers. Another word for a consumer is a heterotroph.

Based on what they eat, we can distinguish between 4 types of heterotrophs.

Consumer	Trophic level	Food source
Herbivores	Primary	Plants
Carnivores	Secondary or higher	Animals
Omnivores	All levels	Plants & animals
Detritivores	----------	detritus

A trophic level refers to the organisms position in the food chain. Autotrophs are at the base. Organisms that eat autotrophs are called herbivores or primary consumers. An organism that eats herbivores is a carnivore and a secondary consumer. A carnivore which eats a carnivore which eats a herbivore is a tertiary consumer, and so on.

It is important to note that many animals do not specialize in their diets. Omnivores (such as humans) eat both animals and plants. Further, except for some specialists, most carnivores don't discriminate between herbivorous and carnivorous bugs in their diet. If it's the right size, and moving at the right distance, chances are the frog will eat it.

Flow of Energy and its Utilisation

The diagram shows how both energy and inorganic nutrients flow through the ecosystem. Energy "flows" through the ecosystem in the form of carbon- carbon bonds. When respiration occurs, the carbon-carbon bonds are broken and the carbon is combined with oxygen to form carbon dioxide.

This process releases the energy, which is either used by the organism (to move its muscles, digest food, excrete wastes, think, etc.) or the energy may be lost as heat. The dark arrows represent the movement of this energy. Note that all energy comes from the sun, and that the ultimate fate of all energy in ecosystems is to be lost as heat. Energy does not recycle.

The other component shown in the diagram is the inorganic nutrients. They are inorganic because they do not contain carbon-carbon bonds. These inorganic nutrients include the phosphorous in your teeth, bones, and cellular membranes the nitrogen in your amino acids (the building blocks of protein); and the iron in your blood (to name just a few of the inorganic nutrients).

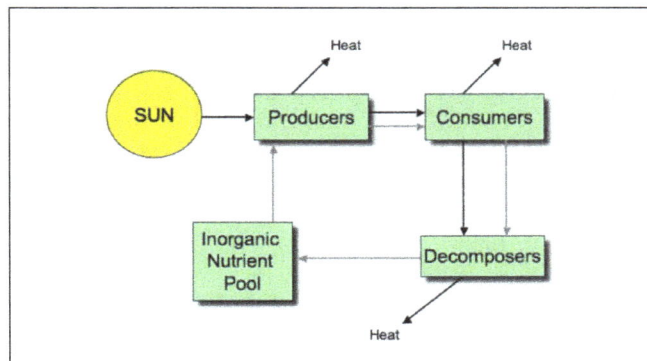

The movement of the inorganic nutrients is represented by the open arrows. Note that the autotrophs obtain these inorganic nutrients from the inorganic nutrient pool, which is usually the soil or water surrounding the plants or algae.

These inorganic nutrients are passed from organism to organism as one organism is consumed by another. Ultimately, all organisms die and become detritus, food for the decomposers. At this stage, the last of the energy is extracted (and lost as heat) and the inorganic nutrients are returned to the soil or water to be taken up again. The inorganic nutrients are recycled, the energy is not.

Characteristics of Energy Flow in an Ecosystem

Unidirectional Flow of Energy

The most important characteristic is the one-way street along which energy flows. The energy that is captured by the autotrophs does not revert back to solar input; that which passes to the herbivores does not pass back to the autotrophs; and so on. The immediate implication of this unidirectional energy flow is that the ecosystem would collapse if the primary energy source, the sun, was cut-off.

Progressive Decrease in Energy

From Figure it is to be noted that at each trophic level there is progressive decrease in energy. This is because at the time of energy transfer from one trophic level to the other a substantial amount of energy is lost as it is dissipated as heat during metabolic activity.

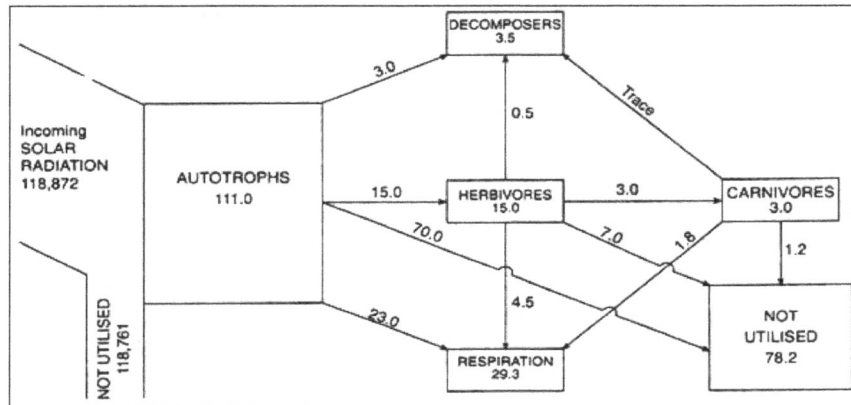

Diagram of energy flow at Ceder Bog Lake, Minnesota in gcal cm^{-2} yr^{-1}.

Respiratory Loss High in Higher Trophic Levels

From the above case of Cedar Bog Lake it can be seen that, as we go higher up the food-chain, respiratory loss gets higher and higher. The respiratory loss in autotrophs is 21 percent, in herbivores 30 percent and in carnivores 60 percent. This higher energy loss in carnivores is due to its greater locomotory activity.

Higher Efficiency of Assimilation at Higher Trophic Level

As we go higher up the trophic level there is greater efficiency of energy assimilation. In the above example of Cedar Bog Lake, the efficiency of solar energy capture by the autotrophs is only 0.1 percent. From the autotrophs to herbivores it is 3.4 percent, and from herbivores to carnivores it is 28.6 percent.

Unutilised Energy

In all ecosystems, despite the utilisation of energy in various metabolisms by different organisms, large amount of energy always remains in the system as standing crop. This indicates that the ecosystem is under-grazed.

Energy Flow

Energy flow in an ecosystem follows the first and second laws of thermodynamics. The energy flow through any trophic level equals the total assimilation at that level, which in turn, equals the production of biomass plus respiratory loss.

The Cycling

The cycling of mineral elements has been included in the energy flow diagram. Unlike energy, nutrients are typically regenerated and retained within the system.

Food Chain

In ecology, food chain is the sequence of transfers of matter and energy in the form of food from organism to organism. Food chains intertwine locally into a food web because most organisms consume more than one type of animal or plant. Plants, which convert solar energy to food by photosynthesis, are the primary food source. In a predator chain, a plant-eating animal is eaten by a flesh-eating animal. In a parasite chain, a smaller organism consumes part of a larger host and may it self be parasitized by even smaller organisms. In a saprophytic chain, microorganisms live on dead organic matter.

Because energy, in the form of heat, is lost at each step, or trophic level, chains do not normally encompass more than four or five trophic levels. People can increase the total food supply by cutting out one step in the food chain: instead of consuming animals that eat cereal grains, the people themselves consume the grains. Because the food chain is made shorter, the total amount of energy available to the final consumers is increased.

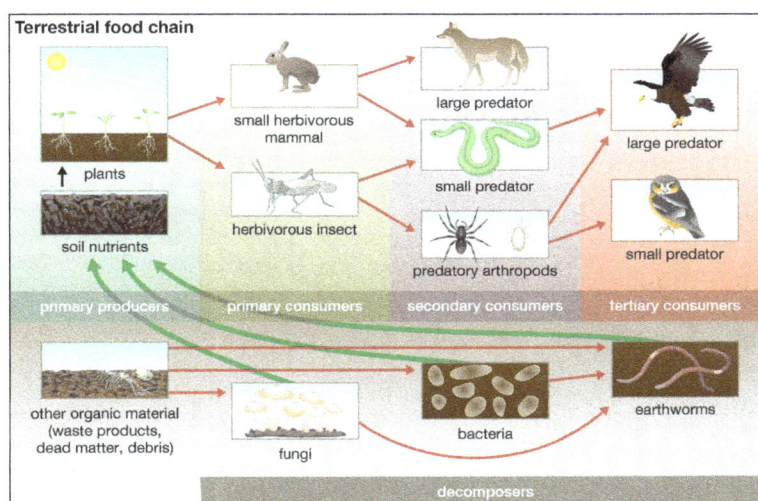

Terrestrial food chain: The terrestrial food chain featuring producers, consumers, and decomposers.

Examples of Food Chains

Food Chains on Land

Land-based food chains represent the most familiar forms of nature to humans. Everything ultimately derives its energy from the sun, and most food chains follow the pattern "herbivore, carnivore, maybe another carnivore or two, apex predator." But there is an almost endless diversity within that pattern and even a few chains that break it.

One fascinating break in that pattern is the omnipresent decomposer. Organisms ranging from bacteria and maggots to the noble cockroach feed on the dead, and in doing so break them down into the nutrients that keep the food chain going. By eating and excreting, decomposers return the nutrients of dead organisms to the soil, which nourishes the plants that start the chains all over again.

- Nectar (flowers) - butterflies - small birds - foxes
- Dandelions - snail - frog - bird - fox
- Dead plants - centipede - robin - raccoon
- Decayed plants - worms - birds - eagles
- Fruits - tapir - jaguar
- Fruits - monkeys - monkey-eating eagle
- Grass - antelope - tiger - vulture
- Grass - cow - man - maggot
- Grass - grasshopper - frog - snake - eagle
- Hazel tree - wood mouse - tawny owl
- Herring - salmon - bear
- Juniper berries - rabbit - fox
- Leaves - ants - anteaters
- Leaves - caterpillars - birds - snakes
- Leaves - giraffes - lions - jackals
- Nuts - squirrels - hawks
- Plants - mice - badgers - bobcats
- Plants - mule deer - mountain lion
- Rice - rat - owl
- Sun - berries - bear - bacteria
- Sun - grass - ant - echidna - dingo
- Sagebrush - elk - wolf
- Switchgrass - earthworm - quail - hawk
- Willow shoots - musk oxen – wolves.

Food Chains in Water

Aquatic food chains are where things get interesting. Much of the ocean remains unexplored, and food chains in water-based environments are often complex and surprising to us land-dwellers. The most famous example is chemosynthesis, But even the aquatic food chains that follow expected patterns can be fascinating.

Decomposers play a crucial role here too, as aquatic decomposers distribute nutrients not just into the soil, but throughout the water column, feeding the plankton that form the base of all aquatic food chains.

- Algae - otocinclus catfish - osprey

- Algae - mosquito larvae - dragonfly larvae - fish - raccoons

- Crayfish - catfish - humans

- Insect - fish - humans

- Mayflies - trout - humans

- Phytoplankton - copepod - fish - squid - seal - orca - brittle star

- Phytoplankton - copepod - bluefish - swordfish - human

- Phytoplankton - copepod - bluehead wrasse - striper - sea cucumber

- Phytoplankton - zooplankton - anchovy - tuna - humans

- Phytoplankton - zooplankton - fish - seal - great white shark

- Phytoplankton - zooplankton - herring - harbor seal

- Plankton - shrimp - herring - cat

- Plankton - snail - mackerel - shark

- Plankton - snail - tuna - dolphin

- Plankton - threadfin shad - bass - humans

- Seaweed - periwinkle - ragworm - curlew

- Caterpillars - turtles - alligators - humans

- Watercress - mayfly larva – stickleback.

Chemosynthetic Food Chains

Every food chain was based on plants turning sunlight into energy. Then, deep-sea submersibles discovered whole ecosystems that existed in the darkest depths of the ocean.

There, microbes that never saw the sun derived nutrients from compounds vented into the water from deep in the Earth's crust and produced chemicals that supported whole new food webs never dreamt of on the surface. That's chemosynthesis. Here are some examples.

- Bacteria - clams - octopus

- Bacteria - copepods - shrimp - zoarcid fish

- Bacteria - tubeworms - zoarcid fish

- Microbes - ridgeia tubeworms - spider crab - octopus

- Microbes - shrimp - crabs

- Mussels - brachyuran crabs - octopus

- Mussels - shrimp - anemone

- Tubeworms - crabs - shrimp - zoarcid fish.

Levels of the Food Chain

The food chain starts from the base, which consists of producer organism and moves up the series in a straight line. The successive levels in a food chain are known as trophic levels. The trophic level of any living organism is determined by the position it occupies within a food chain. There are various trophic levels in a food chain.

Primary Producers

In the food chain, the primary or base level is formed by the autotrophs. These are organisms that are capable of producing their own food from such substance as carbon dioxide and turning it into energy with the help of sunlight. These are plants and algae. They do not consume other organisms, but draw nutrients from the soil or the ocean and by the process of photosynthesis manufacture their own food. They are called primary producers. The energy for the base of the food chain comes from the sunlight.

Primary Consumers

Heterotrophic organisms or those who feed on the first trophic level, the autotroph biomass, form the second trophic level. They are the herbivores and include the tiny crustacean zooplankton that feed on the microscopic algal cells from the surface waters of lakes, ponds, and oceans, as well as much larger, mammalian herbivores, such as mice, deer, cows, and elephants. Herbivores utilize the fixed energy and nutrients in their food of autotrophic biomass to drive their own metabolic processes and to achieve their own growth.

Secondary Consumers

The next link in the chain is animals that eat herbivores – these are called secondary consumers — an example is a snake that eat rabbits. So, snakes are secondary consumers.

Tertiary Consumers

Tertiary consumers are the next level of consumers. Carnivores that consume other carnivores are called tertiary consumers like Killer whales. Killer whales hunt seals and sea lions. These are carnivores that kill fish, squid, and octopuses.

Quaternary Consumers

The next level in the food chain is occupied by the quaternary consumers. They are typically carnivorous animals that eat tertiary consumers. Also known as apex predator or alpha predator or apical predator, they are predator residing at the top of a food chain upon which no other creatures prey.

Decomposers

Decomposers break down dead plants and animals. They also break down the waste of other organisms. They are an important element in the food chain because they keep up a continuous flow of nutrients for the primary producers. Without the decomposers the plants would be unable to source their energy on the one hand and on the other, the environment would be filled with dead matter and waste.

Importance of the Food Chain

The food chains are the living components of the biosphere. These are the vehicles of transfer of energy from one level to another. Through the food chains, transfer of materials and nutrients also takes place.

So a food chain is a picture of organisms in an ecological community that are linked to each other through the transfer of energy and nutrients, beginning with an autotrophic organism such as a plant and continuing with each organism being consumed by one higher in the chain. A food chain also shows how the organisms are related with each other by the food they eat.

We all depend upon food to survive. Energy is necessary for the biotic world to grow and sustain. A food chain describes the method in which a particular organism collects its food. It is an arrangement of who eats whom in a biological community or an ecosystem to obtain food. A food chain is a way of depicting the flow of energy from one organism to the next and to the next and so on. Everyone needs the energy transmitted through a food chain in order to survive.

Components of a Food Chain

The components of a food chain can be represented as,

- Plants – 'base' of the food chain
- Herbivores – feed on plants; many are adapted to live on a diet high in cellulose
- Omnivores – feed on both plants and animals
- Carnivores – feed on herbivores, omnivores, & other carnivores
- 1st level carnivore – feeds on herbivores
- 2nd level carnivore – feeds on 1st level carnivores.

Decomposers

This forms the 'final' consumer group and the most important one. They use energy available in dead plants and animals breaking them down into useful nutrients. The phrase 'food chain' is a way of indicating how energy moves through an ecosystem from the primary producer the green plants to the final consumer the decomposers.

Types of Food Chain

Food is the fundamental source of energy for all living organisms in the ecosystem. A series or chain of organisms where each of them is determined by one and other as a source of energy or food is called a food chain. The food chain is further be divided into two major types; Grazing Food Chain and Detritus Food Chain. The main difference between both these kinds of food chain is that grazing food chain begins from the green plants, which are the principal producers, whereas detritus food chain starts from the dead organic matter or decomposed material that is usually within the soil. The energy to the grazing food chain comes in sunlight as the autotrophs (green plants) prepare their food (photosynthesis) amid the existence of the sunlight. While the energy for the detritus food chain is taken in the detritus or the decomposed materials.

Grazing Food Chain

The grazing food chain is one of the significant varieties of the food chain that's viewed as a food chain procedure dominantly happening in organisms. The grazing food chain begins from the autotrophs (green plants), the significant energy for this series is taken in sunlight as the plants carry out the process of photosynthesis in the presence of sunlight. The green plants operate as the principal producer of the sort of food chain; after the herbivores become fed upon the green plants. The chain goes on farther as the primary consumers (herbivores) are consumed by the secondary consumers (omnivores) in this kind of food chain. This food chain does not consist of germs or other decomposers; it's carried out from the macroscopic organisms. The grazing food chain is the easier kind of food chain as it starts from the main producers (green plants), who are the dominant producers in various ecosystems throughout the planet. The name of the food chain itself informs it have the green plants as a significant source or the one starting off the series.

Detritus Food Chain

The detritus food chain is the sort of food chain which ensures maximum usage and minimal wastage of the available material. This food chain begins from the dead organic matter or other similar wastes; farther, this material is consumed by the creature, and later this creature gets eaten by another animal from the soil. The series keeps on going until the organic matter consists. This sort of food chain is very useful in fixation of inorganic nutrients and using to the maximum. Detritus food chain has the remains of detritus as the significant source of energy, and this procedure gets completed by the subsoil organisms, which could be macroscopic or microscopic. Unlike the grazing food chain, detritus food chain generates a massive quantity of energy to the air.

Comparison Chart

Basis	Grazing Food Chain	Detritus Food Chain
Definition	The grazing food chain begins in the autotrophs (green plants).	Detritus food chain starts from the detritivores.

Energy Supply	In grazing food chain the energy is taken from the sunlight as green plants prepare food in the presence of it.	In detritus food chain the main energy source is stay of detritus.
Organisms	In grazing food chain macroscopic organisms are included.	In detritus food chain subsoil organisms are involved, that could be macroscopic or microscopic.
Number of Energy	Produces a less quantity of energy into the air.	Produces a great quantity of energy to the air.

Energy Transfer in Food Chain

Energy transfer in the food chain is from one trophic level to next. However, not all the energy is transferred as some of it is used for movement, growth, or reproduction repair. The energy flow begins from the primary producers that make their own food by using the solar energy through photosynthesis. In photosynthesis process, the solar energy is converted into chemical energy which is partly used by the plants and the rest stored as carbon compounds.

In any food chain, energy is lost each time one organism eats another. It is therefore necessary, that we have more plants than plant eaters. There have to be more autotrophs than hetrotrophs. The animals in the food chain are interdependent on each other. Each time an organism goes extinct, it disrupts the entire food chain that can have unpredictable consequences.

The energy is then taken up by primary consumers that include the herbivores and omnivores. These organisms acquire the energy by ingesting the plant materials and digesting it in their system to assimilate the stored energy into their bodies.

When secondary consumers eat the primary consumers that have acquired the stored energy from primary consumers, they acquire the assimilated energy. This means the energy is transferred from primary consumers to secondary consumers. As animals in the higher trophic levels such as tertiary and quaternary consumers eat other lower animals for food, the energy is successfully transferred to them.

The energy transfer efficiency through the trophic levels is about 10% and not 100% as the rest is lost in growth, movement, waste, or respiration. So, it means that as the food moves further through the trophic levels, the energy transferred becomes less and less since most of it is lost.

Food Web

Food chains are not isolated sequence, rather, they are interconnected. Most ecosystems contain a number of interconnected trophic interactions, which taken together are referred to as food web. A typical food web for a terrestrial ecosystem is given in figure. As can be seen, a number of food chains operate which ultimately end either at the tertiary consumer level (lion) or at the secondary consumer level (hawk or owl).

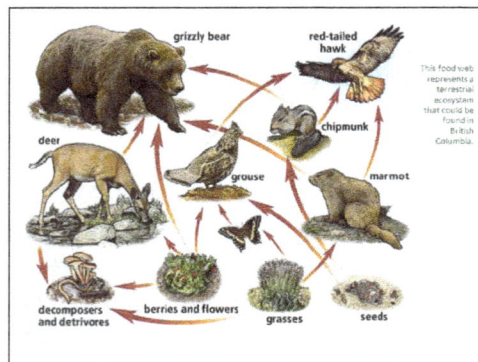

Food web of a terrestrial ecosystem.

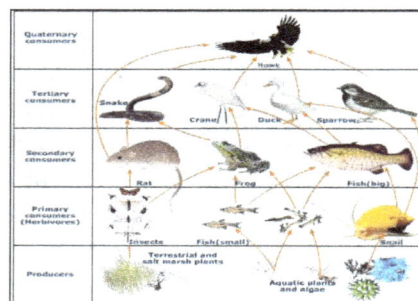

Food web structure in a stocking pond at Kalyani (D_c = detritus consumers, P_p = Primary producer, P_c = primary consumers, S_c = secondary consumers).

Types of Food Web

Three different concepts of food web has been observed by Paine.

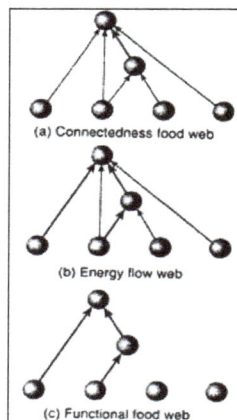

Types of food webs (The heavier the arrow the grater the energy flux).

1. Connectedness food web:

They are also known as topological food web. Such food web emphasises feeding relationships among organisms, portrayed as links in the web. It depicts only the presence or absence of a trophic interaction. They, however, do not show the strength of the interaction, nor any change in trophic relationships. For the above reasons, topological food webs are sometimes referred to as static food web.

2. Energy flow web:

It is sometimes referred to as flow web or as bio-energetic web. It represents an ecosystem viewpoint. Here connections between populations are quantified by the flux of energy between a resource and its consumer.

3. Functional food web:

Also known as interaction food web that identify the feeding relationships within the topological food web that are most important to community structure. The above three food webs depict the importance of each population in maintaining the integrity of a community. Such interactions are the focus of dynamic food web studies.

Reward Feedbacks in Food Webs

Food webs have been found to involve much more than "who eats whom" (predator prey relationship) — an effect known as reward feedback. It has been seen that a "downstream" organism has a positive effect on its "upstream" food supply in the sense that a consumer organism (herbivore) does something that sustains the survival of its food resource (plant).

For example, the fiddler crabs, which feed on surface algae and detritus in coastal marshes, "cultivate" their food plants in several ways. Similarly, vast herds of antelopes grazing on the plains increase the net production of grasses.

In other words, the annual vegetative growth is greater with the grazers than without them. So food webs exhibit partnerships and mutually beneficial relationships between producers and consumers and between different levels of consumers.

Concentration of Toxic Substances along Food Chain

Sometimes certain toxic substances, instead of dispersing, gets concentrated at each link in the chain and are thus referred to as food chain concentration, or, in particular, biological magnification or bioaccumulation. These materials may be put into the environment dilute but may come back concentrated.

Two factors are involved. One, the material remains in the body of the organism and the other is the structure of food chains and trophic levels. An example of such a build-up can be illustrated in the case of DDT (dichlorodiphenyltrichloroethane), a contact insecticide used for the control of mosquitoes. DDT is soluble in fat but not in water.

This toxic insecticide get washed away in various water bodies and get absorbed on detritus. Consequently, when DDT is in or on the food eaten by the herbivore, it tends to be stored in the fat of the animal, rather than excreted or metabolised or denatured.

Subsequently, when these herbivores are eaten up by the carnivores, it results in increased concentration of the insecticide at each step in the food chain, until top predators suffer from very high doses. Prime examples are some of the large seabirds and hawks, where due to high pesticide level, reproductive abnormalities, thin egg shells etc. have been reported.

Bio-magnification may even threaten the human food chain, especially when fishes are affected.

Besides DDT there are DDD, vaclioisotopes and heavy metals like lead, mercury, copper etc. that exhibit bio-magnifications.

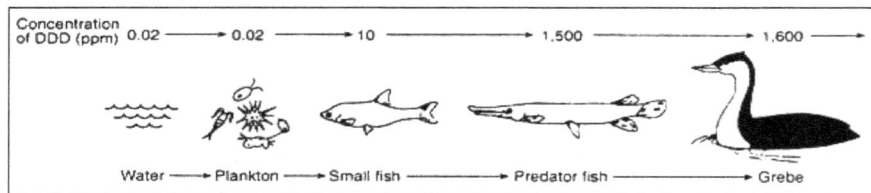

Biomagnification of DDD in Clear Lake.

Applications of Food Webs

1. Food webs are constructed to describe species interactions (direct relationships):

The fundamental purpose of food webs is to describe feeding relationship among species in a community. Food webs can be constructed to describe the species interactions. All species in the food webs can be distinguished into basal species (autotrophs, such as plants), intermediate species (herbivores and intermediate level carnivores, such as grasshopper and scorpion) or top predators (high level carnivores such as fox).

These feeding groups are referred as trophic levels. Basal species occupy the lowest trophic level as primary producer. They convert inorganic chemical and use solar energy to generate chemical energy. The second trophic level consists of herbivores. These are first consumers. The remaining trophic levels include carnivores that consume animals at trophic levels below them. The second consumers (trophic level 3) in the desert food web include birds and scorpions, and tertiary consumers making up the fourth trophic level include bird predators and foxes. Grouping all species into different functional groups or tropic levels helps us simplify and understand the relationships among these species.

2. Food webs can be used to illustrate indirect interactions among species:

Indirect interaction occurs when two species do not interact with each other directly, but influenced by a third species. Species can influence one another in many different ways. One example is the keystone predations are demonstrated by Robert Paine in an experiment conducted in the rocky intertidal zone. This study showed that predation can influence the competition among species in a food web. The intertidal zone is home to a variety of mussels, barnacles, limpets, and chitons. All these invertebrate herbivores are preyed upon by the predator starfish *Piaster*. Starfish was relatively uncommon in the intertidal zone, and considered less important in the community. When Paine manually removed the starfish from experimental plots while leaving other areas undisturbed as control plots, he found that the number of prey species in the experimental plots dropped from 15 at the beginning of the experiment to 8 (a loss of 7 species) two years after the starfish removal while the total of prey species remained the same in the control plots. He reasoned that in the absence of the predator starfish, several of the mussel and barnacle species (that were superior competitors) excluded the other species and reduced overall diversity in the community.

Predation by starfish reduced the abundance of mussel and opened up space for other species to colonize and persist. This type of indirect interaction is called keystone predation.

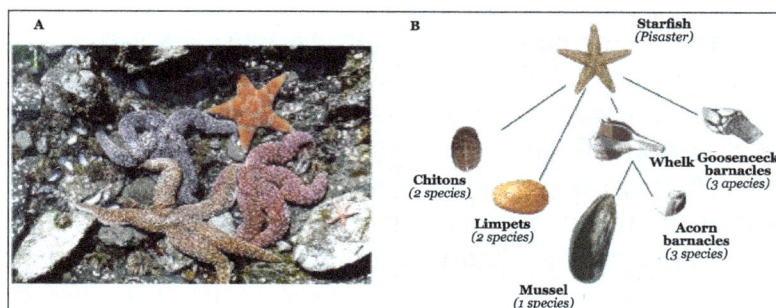

In figure, (a) The rocky intertidal zone of the Pacific Northwest coast is inhabited by a variety of species including tarfish, barnacles, limpets, chitons, and mussels. (b) A food web of this community shows that the starfish preys on a variety of invertebrate species. Removal of starfish from this community reduced the diversity of prey species due to increased competition.

3. Food webs can be used to study bottom-up or top-down control of community structure:

Food webs illustrate energy flow from primary producers to primary consumers (herbivores) and from primary consumers to secondary consumers (carnivores). The structure of food webs suggests that productivity and abundance of populations at any given trophic level are controlled by the productivity and abundance of populations in the trophic level below them. This phenomenon is call bottom-up control. Correlations in abundance or productivity between consumers and their resources are considered as evidence for bottom-up control. For example, plant population densities control the abundance of herbivore populations which in turn control the densities of the carnivore populations. Thus, the biomass of herbivores usually increases with primary productivity in terrestrial ecosystems.

Top-down control occurs when the population density of a consumer can control that of its resource, for example, predator populations can control the abundance of prey species. Under top-down control, the abundance or biomass of lower trophic levels depends on effects from consumers at higher trophic levels. A trophic cascade is a type of top-down interaction that describes the indirect effects of predators. In a trophic cascade, predators induce effects that cascade down the food chain and affect biomass of organisms at least two links away. Nelson Hairston, Frederick Smith and Larry Slobodkin first introduced the concept of top-down control with the frequently quoted "the world is green" proposition. They proposed that the world is green because carnivores depress herbivores and keep herbivore populations in check. Otherwise, herbivores would consume most of the vegetation. Indeed, a bird exclusion study demonstrated that there were significantly more insects and leaf damage in plots without birds compared to the control.

4. Food webs can be used to reveal different patterns of energy transfer in terrestrial and aquatic ecosystems:

Patterns of energy flow through different ecosystems may differ markedly in terrestrial and aquatic ecosystems. Food webs (i.e., energy flow webs) can be used to reveal these differences. In a review paper, Shurin et al. provided evidence for systematic difference in energy flow and biomass partitioning between producers and herbivores, detritus and decomposers, and higher trophic levels in food webs. A dataset synthesized by Cebrian and colleagues on the fate of carbon fixed by primary productivity across different ecosystems was used to show different patterns in food chains between terrestrial and aquatic ecosystems. On average, the turnover rate of phytoplankton is 10 to 1000 times faster than that of grasslands and forests, thus, less carbon is stored in the living autotroph biomass pool, and producer biomass is consumed by aquatic herbivores at 4 times the terrestrial rate. Herbivores in terrestrial ecosystems are less abundant but decomposers are much more abundant than in phytoplankton dominated aquatic ecosystems. In most terrestrial ecosystems with high standing biomass and relatively low harvest of primary production by herbivores, the detrital food chain is dominant. In deep-water aquatic ecosystems, with their low standing biomass, rapid turnover of organisms, and high rate of harvest, the grazing food chain may be dominant.

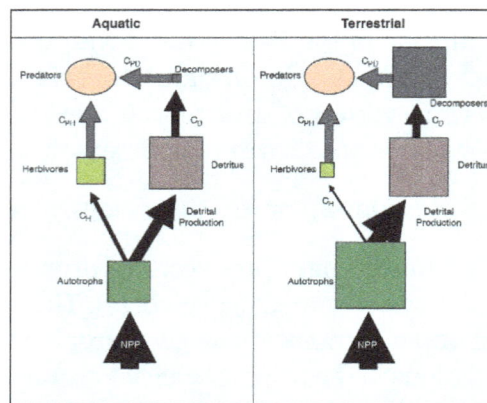

Differences in pathways of carbon flow and pools between
aquatic and terrestrial ecosystems.

The thickness of the arrows (flows) and the area of the boxes (pools) correspond to the magnitude. The size of the pools are scaled as log units since the differences cover four orders of magnitude. The C's indicate consumption terms (i.e. CH is consumption by herbivores). Ovals and arrows in grey indicate unknown quantities.

As a diagram tool, food web has been approved to be effective in illustrating species interactions and testing research hypotheses. It will continue to be very helpful for us to understand the associations of species richness/diversity with food web complexity, ecosystem productivity, and stability.

Ecological Pyramid

An ecological pyramid is a graphical representation of the relationship between different organisms in an ecosystem. Each of the bars that make up the pyramid represents a different trophic level, and their order, which is based on who eats whom, represents the flow of energy. Energy moves up the pyramid, starting with the primary producers, or autotrophs, such as plants and algae at the very bottom, followed by the primary consumers, which feed on these plants, then secondary consumers, which feed on the primary consumers, and so on. The height of the bars should all be the same, but the width of each bar is based on the quantity of the aspect being measured.

Ecological Pyramid Examples

The diagram below is an example of a productivity pyramid, otherwise called an energy pyramid. The sun has been included in this diagram as it's the main source of all energy, as well the *decomposers*, like bacteria and fungi, which can acquire nutrients and energy from all trophic levels by breaking down dead or decaying organisms. As shown, the nutrients then go back into the soil and are taken up by plants.

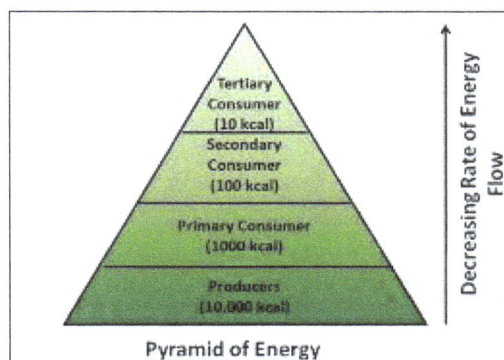

Pyramid of Energy

The loss of energy to the surroundings is also shown in this diagram, and the total energy transfer has been calculated. We start off with the total amount of energy that the primary producers contain, which is indicated by 100%. As we go up one level, 90% of that energy is used in ways other than to create flesh. What the primary consumers end up with is just 10% of the starting energy, and, 10% of that 10% is lost in the transfer to the next level. That's 1%, and so on. The predators at the apex, then, will only receive 0.01% of the starting energy. This inefficiency in the system is the reason why productivity pyramids are always upright.

Function of Ecological Pyramid

An ecological pyramid not only shows us the feeding patterns of organisms in different ecosystems, but can also give us an insight into how inefficient energy transfer is, and show the influence that a change in numbers at one trophic level can have on the trophic levels above and below it. Also, when data are collected over the years, the effects of the changes that take place in the environment on the organisms can be studied by comparing the data. If an ecosystem's conditions are found to be worsening over the years because of pollution or overhunting by humans, action can be taken to prevent further damage and possibly reverse some of the present damage.

Types of Ecological Pyramids

Pyramid of Biomass

A pyramid of biomass is a graphical representation of biomass present in a unit area of various trophic levels. It shows the relationship between biomass and trophic level quantifying the biomass available in each trophic level of an energy community at a given time.

There are two main types of biomass pyramid – inverted pyramid of biomass and the upright one. A good example of the inverted pyramid is in a pond ecosystem where the mass of phytoplankton, the major producers, will always be lower than the mass of the heterotrophs like fish and insects.

As the value of biomass become larger, the pyramid gains an inverted shape with tertiary consumers appearing at the top in biomass.

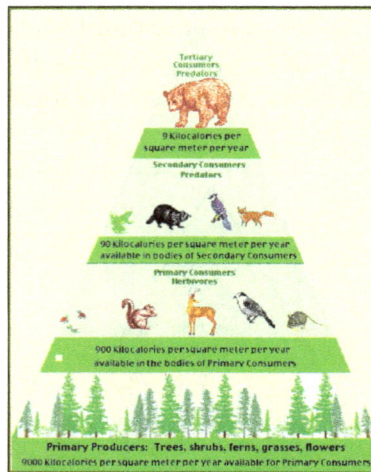

Biomass

In ecological terms, biomass refers to the total mass of all living or organic matter that inhabit an ecosystem at any given point of time. There are two main different types of biomass:

- Species Biomass

- Community Biomass

Species biomass is the total mass of species in an ecosystem. Community biomass, on the other hand, is the total mass of all the species that regard the specified community as their habitat. When it comes to measuring biomass, the species can include human beings and even microorganisms.

The method used in measuring biomass depends on the reason for which the biomass is being measured. You may want to take the mass of the organisms in their natural state and their natural habitat for biomass calculation. Say you want to calculate the biomass in a fishery, for example:

The total biomass should be the biomass of the fish while they are still wet if you took them out of the water. Conversely, if the dried mass of the fish is taken, it would account for 30 percent of their actual mass. This is because the rest of the mass will be water. Measurement of biomass in terms of dry weight is more accurate.

In some other cases, the mass would only include biological tissues. Bone mass, teeth mass, or shell mass, wouldn't be considered. This case applies where you want to measure only the carbon present in the body.

Biomass is an expression of the mass per unit area. Hence, the units of measurements are grams per square meter or tons per square kilo meter.

Significance of Pyramid of Biomass

A biomass pyramid is useful for quantifying the biomass that is available as a result of organisms at every trophic level. This pyramid begins with the producer, normally the plants, which occupy the bottom level of the pyramid. The producers are followed by primary consumers.

The highest quantified amount of biomass sits at the topmost level of the pyramid. This level largely includes the carnivores. Note that we are talking about an upright pyramid here.

Virtually all of the world's ecosystems and biomes are represented by an upright biomass pyramid. In an upright pyramid ecosystem, the total weight of the producers is more than the total weight of the consumers. However, the inverted pyramid of biomass is the complete opposite. Namely, the combined weight of the producers is less than the combined weight of the consumers.

The manner in which biomass pyramid is represented is based on the law of thermodynamics. This law states that energy can never be destroyed. It can never be created either. It can only be transformed from one form into another. Namely, the energy is transferred through the chain from producers to consumers and so on, and converted in the biomass.

Major Limitation of a Biomass Pyramid

One of the main limitations of a biomass pyramid is that every trophic level seems to have more energy than it truly does.

A good example to illustrate this is when human beings consume another animal. The mass of the animal's bones is calculated. However, the mass of the bones is not actually utilized in the next level of the pyramid of biomass.

A biomass pyramid counts mass that is not actually transmitted to the next trophic level. Nonetheless, a pyramid of biomass remains one of the excellent ways to determine if there is an imbalance in an ecosystem.

Examples of Biomass Pyramid

We have an inverted biomass pyramid and the upright one. Examples of the normal biomass pyramid include:

- Mice eat grass seeds. The mice are in turn eaten by the owl. The grass has the greatest biomass in this chain. Its biomass, therefore, sits at the bottom of the pyramid. Conversely, the owl has the lowest biomass in the chain and hence sits on top of the pyramid.

- A caterpillar feeds on an oak tree. A caterpillar is in turn eaten by a blue tit, which is eaten by a Sparrowhawk. The oak tree sits at the bottom of the biomass pyramid as it can feed dozens of caterpillars, thanks to its massive biomass. The Sparrowhawk occupies the highest level of the pyramid.

Pyramid of Numbers

A pyramid of numbers is a graphical representation that shows the number of organisms at each trophic level. It is an upright pyramid in light of the fact that in an ecosystem, the producers are always more in number than other trophic levels.

The pyramid of numbers was advanced by Charles Elton in 1927. Charles pointed out the huge difference in the number of organisms involved in each level of the food chain. Succeeding links of the trophic structure reduce rapidly in number until there's a very small number of carnivores at the top.

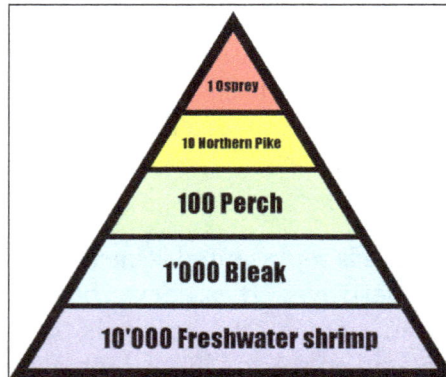

The reason for the pyramid shape is that there must always be enough plants to produce food at the bottom. Otherwise, the entire food chain would collapse. At the higher level, no predator can be as common as its prey. Otherwise, the population of both animals would soon be wiped out. Sparrowhawks, for example, can never be common than blue tits if they live in the same ecosystem.

This pyramid doesn't consider the biomass of organisms. Neither does it indicate the energy transmitted or the utilization of energy by the organisms involved. Numbers pyramid can be convenient since counting is usually a simple task and can be undertaken over the years to track the changes in a given ecosystem.

The pyramid can also be used to figure out how the population of a given species can affect another. Additionally, it can serve as a basis for an ecosystem's quantitative analysis.

But it is worth noting that some types of organisms can be difficult to count, especially in the case of some juvenile forms. The unit of measurement in the pyramid of numbers is the number of organisms. This pyramid varies between ecosystems and is of three types.

Types of Pyramid of Numbers

1. Upright pyramid of number.

2. Partly upright pyramid of number.

3. The inverted pyramid of number.

1. Upright Pyramid of Number:

This type of number pyramid is found in the grassland ecosystem. This ecosystem is characterized by numerous autotrophs that support lesser herbivores. The herbivores, in turn, support a smaller number of carnivores.

Therefore, this pyramid is upright. Namely, with every higher trophic level, the number of organisms decreases.

For example, the grasses sit at the lowest trophic level or the base of the number pyramid because of their abundance. The primary consumer, such as a grasshopper, occupies the next higher trophic level. Grasshoppers are fewer in number than grass. The next trophic level is a primary carnivore, such as a rat.

There are fewer rats than grasshoppers because they consume grasshoppers. Secondary carnivores, such as snakes, occupy the next higher trophic level. Snakes feed on rats and snakes are eaten by hawks, which occupy the highest trophic level and are the least in number.

A pond ecosystem also depicts an upright pyramid of numbers. Phytoplankton like algae and bacteria are the producers here and hence the highest in number. The smaller herbivorous fishes are fewer in number compared to producers.

Likewise, the small carnivorous fishes are less in number than the herbivorous ones. Lastly, the apex consumers or largest carnivorous fishes are the least in number.

2. Partly Upright Pyramid of Number:

This type of number pyramid is typical of the forest ecosystem. In this ecosystem, the producers are large-sized trees, which sit at the base of the number pyramid. The herbivores, such as elephants and fruit-eating birds, make the primary consumers. They are more in number than the producers. Afterward, the number of individual organisms reduces at each successive trophic level.

3. Inverted Pyramid of Number:

An inverted number pyramid is found in parasitic food chains. In these food chains, there's normally one producer supporting numerous parasites. The parasites, in turn, support more hyper-parasites. In short, in this pyramid, number of individuals at each level is increased from lower level to higher level.

Examples of Pyramid of Number

1. Clover → Snail → Thrush → Hawk

Being a plant, clover is the producer of this food chain and hence sits at the bottom of the pyramid. Energy is lost to the environment as you go up from one level to the next. That means there are fewer organisms at each step in this food chain. There is a need for plenty of clovers to support the snail population.

A thrush feeds on plenty of snails. A hawk, in turn, eats plenty of thrushes. Hence, the population of hawks is very small.

2. Phytoplankton → Zooplankton → Small Crustaceans → Predator Insects → Small Fish → Large Fish → Kingfisher

In this aquatic food chain, the phytoplankton is consumed by the zooplankton, which is in turn eaten by the small crustaceans. Then the predator insects feed on the small crustaceans. The predator insects are in turn consumed by the small fish, which are eaten by the large fish. Lastly, the large fish are eaten by the kingfisher. The Kingfisher is the least in number in the food chain and sits at the apex of the pyramid.

In some cases, the pyramid may not look like a pyramid at all. That's why we have a partly upright pyramid of numbers. This normally happens if the producer is a large plant or if one of the animals is very small. Whatever the situation, however, the producer still occupies the bottom of the pyramid.

Some examples of this case include:

3. Oaktree → Caterpillars → Blue Tit → Sparrowhawk

A large number of caterpillars can feed on a single oak tree. The caterpillars provide food for several blue tits that in turn are eaten by a sparrowhawk.

4. Oaktree → Insects → Woodpecker

Many insects can feed on an oak tree since it is very large.

5. Grass → Rabbit → Flea

Fleas are very small and plenty of them can feed on one rabbit.

Energy Pyramid

Energy Pyramid is sometimes referred to as an ecological pyramid or trophic pyramid. It is a graphical representation between various organisms in an ecosystem. The pyramid is composed of several bars. Each bar has a different trophic level to represent.

The order of these bars is based on who feeds on whom. It represents the energy flow in the ecosystem. Energy flows from the bottom of the pyramid, where we have producers, upwards. The height of the bars is normally the same. However, each bar has a different width depending on the quantity of the element being measured.

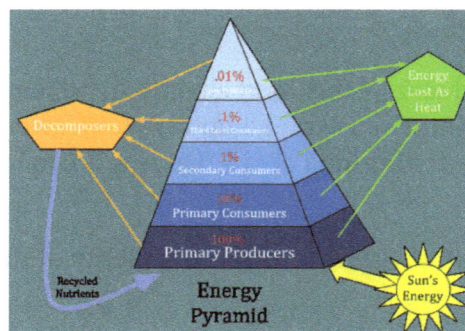

An energy pyramid is useful in quantifying the transfer of energy from one organism to another along a food chain. Energy is higher at the bottom of the pyramid, but it decreases as you move up through the trophic levels.

Namely, as energy flows through the various trophic levels, some energy is normally dissipated as heat at each level. About 10% of the total energy is transferred during energy flow through several trophic levels and hence the steady drop in the amount of energy.

The shape is significant in demonstrating the flow of energy due to the way the energy is utilized and lost throughout the ecosystem.

Four Main Levels of Energy Pyramid

1. Producers

The producers and the energy available within them occupy the first level of the energy pyramid. These producers are largely the autotrophs – organisms that manufacture their own food by harnessing energy from non-living sources of energy. Often times, these are photosynthesizing plants.

These plants use solar energy to manufacture their own food in the form of simple sugars. Some autotrophs don't get their energy from the sun directly but from the soil. These autotrophs include earthworms and fungi like mushrooms.

However, the energy that producers like mushrooms and earthworms receive from the soil is less than what green plants get from the sun. Namely, the energy from the soil experiences an additional layer of filtering through the soil. Therefore, a robin that feeds on a worm, for example, would get less energy than it would if it feeds on a berry instead.

In the other levels of the energy pyramid, we only have heterotrophs – organisms that get their food from organic carbon, normally from other organisms.

2. Primary Consumers

The second level of the energy pyramid is represented by primary consumers, which are usually herbivores. Herbivores are animals that depend only on the plants for their nourishment and survival. After generating their energy from the sun, plants pass the energy on to the primary consumers.

This facilitates the transfer of solar energy from one trophic level to another. Human beings don't fully depend on the primary consumers, but it is imperative that this level is present in the ecosystem. Otherwise, the system won't function normally.

3. Secondary Consumers

Secondary consumers sit on the third level of the energy pyramid. They are commonly known as the carnivores. Secondary consumers are organisms that depend on the primary consumers for their nourishment and survival.

Without the primary consumers, the carnivores wouldn't have anything to eat and hence not exist. In this level, the energy that was given to the primary consumers from the producers is now transmitted to this level. This facilitates the smooth flow of the energy for effective use.

It is worth noting that there are different percentages of energy that are transmitted to various ecosystem levels depending on the amount of energy supplied to the producers (plants).

4. Tertiary Consumers

The last level of the energy pyramid encompasses the tertiary consumers. It is the level of the secondary carnivores that feed on both the primary and the secondary consumers. The energy level of the ecosystem is finished at this level.

The energy that is normally not utilized by the plants goes back to the environment, which includes the soil, the water bodies, and the atmosphere. It is then normally released to the outer space. It is imperative that all the different levels of the energy pyramid get sufficient energy as required to ensure the earth remains stable.

Throughout the whole energy pyramid, the decomposers have a critical role to play. These decomposers, which include bacteria, worm, and fungi, break down the tissues and other organic matter that have not been consumed by the organisms higher in the pyramid. They also use up the little amount of energy that remains in the tissues of dead organisms.

In so doing, these decomposers recycle the nutrients back into the soil, contributing greatly to the carbon and nitrogen cycles.

Examples of Energy Pyramid

There are countless examples of energy pyramid that can help you better understand the concept. Here are three common examples:

1. An earthworm breaks down dead organic matter in the soil which the plants, sitting one level up in the pyramid utilize to manufacture their food along with the light from the sun during the photosynthesis process. The herbivores in the next level up in the pyramid, in turn, use the stored energy in the plants by feeding on the plants. The energy contained in the fecal matter from the herbivores is recycled back into the system where it is broken down further by the earthworms.

2. Mice on the forest floor eat the seeds and fruits of trees, shrubs, and flowers for energy. The eagle, sitting at the next level up the energy pyramid eats the mice, taking in the energy they have stored. It is worth noting that adult eagles have no natural predators. That means they occupy the topmost level of their energy pyramid.

3. Grasshoppers eat grass for their energy. The grasshoppers, in turn, give their energy to frogs in the next level up the pyramid, which feed on them. Snakes in the next level of the pyramid get their energy from frogs and so on.

References

- Food-chain, science: britannica.com, Retrieved 19 August, 2019

- Examples-of-food-chains: yourdictionary.com, Retrieved 11 June, 2019

- Levels-importance-components-food-chain, ecosystem: eartheclipse.com, Retrieved 10 June, 2019

- Grazing-food-chain-vs-detritus-food-chain: diffzi.com, Retrieved 20 February, 2019

- Food-web-meaning-and-types-zoology, food-web: notesonzoology.com, Retrieved 17 July, 2019

- Food-web-concept-and-applications: nature.com, Retrieved 7 May, 2019

- Ecological-pyramid: biologydictionary.net, Retrieved 30 August, 2019

- Pyramid-of-biomass-definition-examples, ecosystem: eartheclipse.com, Retrieved 4 April, 2019

- Pyramid-of-numbers-types-and-examples, ecosystem: eartheclipse.com, Retrieved 1 July, 2019

- Energy-pyramid-definition-levels-examples, ecosystem: eartheclipse.com, Retrieved 19 March, 2019

Chapter 4

Population Ecology

The subfield of ecology which studies the dynamics of species populations along with the way in which they interact with the environment is known as population ecology. There are different ways to understand the dynamics between populations such as predator-prey model and Allee effect. The diverse applications of these theories and models related to population ecology have been thoroughly discussed in this chapter.

Population ecology is the study of the processes that affect the distribution and abundance of animal and plant populations.

A population is a subset of individuals of one species that occupies a particular geographic area and, in sexually reproducing species, interbreeds. The geographic boundaries of a population are easy to establish for some species but more difficult for others. For example, plants or animals occupying islands have a geographic range defined by the perimeter of the island. In contrast, some species are dispersed across vast expanses, and the boundaries of local populations are more difficult to determine. A continuum exists from closed populations that are geographically isolated from, and lack exchange with, other populations of the same species to open populations that show varying degrees of connectedness.

Genetic variation within Local Populations

In sexually reproducing species, each local population contains a distinct combination of genes. As a result, a species is a collection of populations that differ genetically from one another to a greater or lesser degree. These genetic differences manifest themselves as differences among populations in morphology, physiology, behaviour, and life histories; in other words, genetic characteristics (genotype) affect expressed, or observed, characteristics (phenotype). Natural selection initially operates on an individual organismal phenotypic level, favouring or discriminating against individuals based on their expressed characteristics. The gene pool (total aggregate of genes in a population at a certain time) is affected as organisms with phenotypes that are compatible with the environment are more likely to survive for longer periods, during which time they can reproduce more often and pass on more of their genes.

The amount of genetic variation within local populations varies tremendously, and much of the discipline of conservation biology is concerned with maintaining genetic diversity within and among populations of plants and animals. Some small isolated populations of asexual species often have little genetic variation among individuals, whereas large sexual populations often have great variation. Two major factors are responsible for this variety: mode of reproduction and population size.

Effects of Mode of Reproduction: Sexual and Asexual

In sexual populations, genes are recombined in each generation, and new genotypes may result. Offspring in most sexual species inherit half their genes from their mother and half from their father, and their genetic makeup is therefore different from either parent or any other individual in the population. In both sexually and asexually reproducing species, mutations are the single most important source of genetic variation. New favourable mutations that initially appear in separate individuals can be recombined in many ways over time within a sexual population.

In contrast, the offspring of an asexual individual are genetically identical to their parent. The only source of new gene combinations in asexual populations is mutation. Asexual populations accumulate genetic variation only at the rate at which their genes mutate. Favourable mutations arising in different asexual individuals have no way of recombining and eventually appearing together in any one individual, as they do in sexual populations.

Effects of Population Size

Over long periods of time, genetic variation is more easily sustained in large populations than in small populations. Through the effects of random genetic drift, a genetic trait can be lost from a small population relatively quickly. For example, many populations have two or more forms of a gene, which are called alleles. Depending on which allele an individual has inherited, a certain phenotype will be produced. If populations remain small for many generations, they may lose all but one form of each gene by chance alone.

This loss of alleles happens from sampling error. As individuals mate, they exchange genes. Imagine that initially half of the population has one form of a particular gene, and the other half of the population has another form of the gene. By chance, in a small population the exchange of genes could result in all individuals of the next generation having the same allele. The only way for this population to contain a variation of this gene again is through mutation of the gene or immigration of individuals from another population.

Minimizing the loss of genetic variation in small populations is one of the major problems faced by conservation biologists. Environments constantly change, and natural selection continually sorts through the genetic variation found within each population, favouring those individuals with phenotypes best suited for the current environment. Natural selection, therefore, continually works to reduce genetic variation within populations, but populations risk extinction without the genetic variation that allows populations to respond evolutionarily to changes in the physical environment, diseases, predators, and competitors.

Population Density and Growth

Life Histories and the Structure of Populations

An organism's life history is the sequence of events related to survival and reproduction that occur from birth through death. Populations from different parts of the geographic range that a species inhabits may exhibit marked variations in their life histories. The patterns of demographic variation seen within and among populations are referred to as the structure of populations. These variations include breeding frequency, the age at which reproduction begins, the number of times an individual reproduces during its lifetime, the number of offspring produced at each reproductive episode (clutch or litter size), the ratio of male to female offspring produced, and whether reproduction is sexual or asexual. These differences in life history characteristics can have profound effects on the reproductive success of individuals and the dynamics, ecology, and evolution of populations.

Of the many differences in life history that occur among populations, age at the time of first reproduction is one of the most important for understanding the dynamics and evolution of a population. All else being equal, natural selection will favour, within species, individuals that reproduce earlier than other individuals in the population, because by reproducing earlier an individual's genes enter the gene pool (the sum of a population's genetic material at a given time) sooner than those of other individuals that were born at the same time but have not reproduced. Nonetheless, the "all else being equal" qualification is an important one because delayed reproductive strategies that ensure larger and more-robust offspring may be selected for in some species of long-lived organisms. Precocial development (unusually early maturation) to reproduction may be favoured, however, if the genes of early reproducers begin to spread throughout the population. Individuals whose genetic makeup allows them to reproduce earlier in life will come to dominate a population if there is no counterbalancing advantage to those individuals that delay reproduction until later in life.

Not all populations, however, are made up of individuals that reproduce very early in life. In the course of a lifetime, an individual must devote energy and resources to physiological demands other than reproduction. This is referred to as the cost of reproduction. To reproduce successfully, a plant first may have to grow to a certain height and out compete its neighbours, and an animal may have to devote energy to growth so that it can reach a size at which it can fend off predators and successfully compete for mates. In many populations, individuals that delay reproduction have a better chance of surviving and leaving offspring than those that attempt to reproduce early. The opposing demands of growth, defense, and reproduction are balanced within the constraints of different environments to produce populations that have a diverse range of life history strategies.

Populations often can be divided into one of two extreme types based on their life history strategy. Some populations, called r-selected, are considered opportunistic because their reproductive behaviour involves a high intrinsic rate of growth (r)—individuals give birth once at an early age to many offspring. Populations that exhibit this strategy often have been shaped by an extremely variable and uncertain environment. Because mortality occurs randomly in this setting, quantity of progeny rather than quality of care serves the species better. In another strategy, called K-selected, populations tend to remain near the carrying capacity (K), the maximum number of individuals that the environment can sustain. Individuals in a K-selected population give birth at a later age to fewer offspring. This equilibrial life history is exhibited in more stable environments where reproductive success depends more on the fitness of the offspring than on their numbers.

K-selected species: Adult and young African savanna elephants (Loxodonta africana) crossing a stream.

Elephants are classic examples of K-selected species—that is, species characterized by relatively stable populations. Such species produce a few large young instead of many small young.

Life Tables and the Rate of Population Growth

Differences in life history strategies, which include an organism's allocation of its time and resources to reproduction and care of offspring, greatly affect population dynamics. As stated above, populations in which individuals reproduce at an early age have the potential to grow much faster than populations in which individuals reproduce later. The effect of the age of first reproduction on population growth can be seen in the life tables for a particular species. Life tables were originally developed by insurance companies to provide a means of determining how long a person of a particular age could be expected to live. They are used not only by demographers of human populations but also by plant, animal, and microbial ecologists to make projections about the life expectancies of nonhuman populations, as well as the effects of variation on demography and population growth. The number of individuals in a closed population (a population in which neither immigration nor emigration occurs) is governed by the rates of birth (natality), growth, reproduction, and death (mortality). Life tables are designed to evaluate how these rates influence the overall growth rate of a population.

Survivorship Curves

Life tables follow the fate of a group of individuals all born within the same population in the same year. Of this group, or cohort, only a certain number of individuals will reach each age, and there is an age above which no individuals ever survive. Plotting the number of those members of the group that are still alive at each age results in a survivorship curve for the population. Survivorship curves are usually displayed on a semi logarithmic rather than an arithmetic scale.

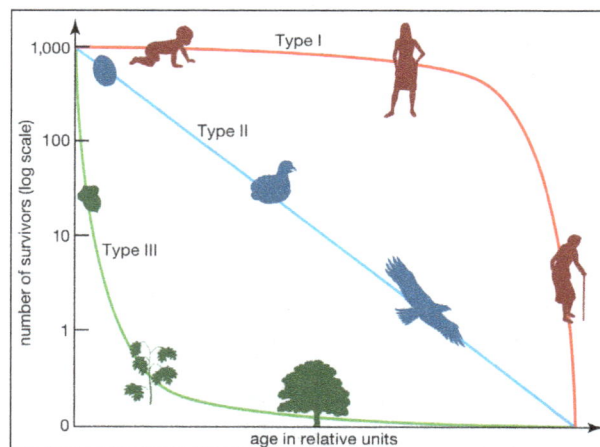

Survivorship curve: Type I, II, and III survivorship curves. A survivorship curve is the graphic representation of the number of individuals in a population that can be expected to survive to any specific age.

There are three general types of survivorship curves. Species such as humans and other large mammals, which have fewer numbers of offspring but invest much time and energy in caring for their young (K-selected species), usually have a Type I survivorship curve. This relatively flat curve reflects low juvenile mortality, with most individuals living to old age. A constant probability of dying at any age, shown by the Type II survivorship curve, is evident as a straight line with a constant slope that decreases over time toward zero. Certain lizards, perching birds, and rodents exhibit this type of survivorship curve. In some species that produce many offspring but provide little care for them (r-selected species), mortality is greatest among the youngest individuals. The Type III survivorship curve indicative of this life history is initially very steep, which is reflective of very high mortality among the young, but flattens out as those individuals who reach maturity survive for a relatively longer time; it is exhibited by animals such as many insects or shellfish. Many populations have survivorship patterns that are more complex than, or fall in between, these three idealized curves. For example, passerine birds (perching birds such as finches) commonly suffer high mortality during the first year of life and a lower, more constant rate of death in subsequent years.

Calculating Population Growth

Life tables also are used to study population growth. The average number of offspring left by a female at each age together with the proportion of individuals surviving to each age can be used to evaluate the rate at which the size of the population changes over time. These rates are used by demographers and population ecologists to estimate population growth and to evaluate the effects of conservation efforts on endangered species.

Galapagos cactus finch (Geospiza scandens): It has such a high reproductive rate that
the population can more than double in size each generation.

The average number of offspring that a female produces during her lifetime is called the net reproductive rate (R_o). If all females survived to the oldest possible age for that population, the net reproductive rate would simply be the sum of the average number of offspring produced by females at each age. In real populations, however, some females die at every age. The net reproductive rate for a set cohort is obtained by multiplying the proportion of females surviving to each age (l_x) by the average number of offspring produced at each age (m_x) and then adding the products from all the age groups: $R_o = \Sigma l_x m_x$. A net reproductive rate of 1.0 indicates that a population is neither increasing nor decreasing but replacing its numbers exactly. This rate indicates population stability. Any number below 1.0 indicates a decrease in population, while any number above indicates an increase. For example, the net reproductive rate for the Galapagos cactus finch (*Geospiza scandens*) is 2.101, which means that the population can more than double its size each generation.

Life table for one Darwin finch, the Galapagos cactus finch (Geospiza scandens)*			
age class** (x)	probability of surviving to age x (l_x)	average number of fledgling daughters (m_x)	product of survival and reproduction ($\Sigma l_x m_x$)

0	1.0	0.0	0.0
1	0.512	0.364	0.186
2	0.279	0.187	0.052
3	0.279	1.438	0.401
4	0.209	0.833	0.174
5	0.209	0.500	0.104
6	0.209	0.833	0.174
7	0.209	0.250	0.052
8	0.209	3.333	0.696
9	0.139	0.125	0.017
10	0.070	0.0	0.0
11	0.070	0.0	0.0
12	0.070	3.500	0.245
13	0	—	—
			R_o = 2.101
Net reproductive rate = R_o = $\Sigma l_x m_x$ = 2.101			
Mean generation time = T = $(\Sigma x l_x m_x)/(R_o)$ = 6.08 years			
Intrinsic rate of natural increase of the population = r = approximately $\ln R_o / T$ = 2.101/6.08 = 0.346			
*The values are for the cohort of females born in 1975.			
**Designated in years.			

The other value needed to calculate the rate at which the population can grow is the mean generation time (T). Generation time is the average interval between the birth of an individual and the birth of its offspring. To determine the mean generation time of a population, the age of the individuals (x) is multiplied by the proportion of females surviving to that age (l_x) and the average number of offspring left by females at that age (m_x). This calculation is performed for each age group, and the values are added together and divided by the net reproductive rate (R_o) to yield the result.

$$T = \frac{\sum x l_x m_x}{R_o}$$

For example, the mean generation time of the Galapagos cactus finch is 6.08 years.

Another value is used by population biologists to calculate the rate of increase in populations that reproduce within discrete time intervals and possess generations that do not overlap. This is known as the intrinsic rate of natural increase (r), or the Malthusian parameter. Very simply, this rate can be understood as the number of births minus the number of deaths per generation time—in other words, the reproduction rate less the death rate. To derive this value using a life table, the natural logarithm of the net reproductive rate is divided by the mean generation time:

$$r \frac{\ln R_o}{T}$$

Values above zero indicate that the population is increasing; the higher the value, the faster the growth rate. The intrinsic rate of natural increase can be used to compare growth rates of populations of a species that have different generation times. Some human populations have higher intrinsic rates of natural increase partially because individuals in those groups begin reproducing earlier than those in other groups. Mice have higher intrinsic rates of natural increase than elephants because they reproduce at a much earlier age and have a much shorter mean generation time.

If a population has an intrinsic rate of natural increase of zero, then it is said to have a stable age distribution and neither grows nor declines in numbers. A growing population has more individuals in the lower age classes than does a stable population, and a declining population has more individuals in the older age classes than does a stable population. Many human populations are currently undergoing population increase, far exceeding a stable age distribution. Although the global human population has increased almost continuously throughout history, it has skyrocketed since the Industrial Revolution, primarily because of a drop in death rates. No other species has shown such sustained growth.

Intrinsic rate of increase (r)* calculated for populations of species that differ greatly in their potential for the rate of population growth	
species	intrinsic rate of increase (r)
Elephant seal	0.091
Ring-necked pheasant	1.02
Field vole	3.18
Flour beetle	23
Water flea	69
*Values above zero indicate that the population is increasing. The higher the value of r, the faster the intrinsic growth rate of the population.	

Limits to Population Growth

Exponential and Geometric Population Growth

In an ideal environment, one that has no limiting factors, populations grow at a geometric rate or an exponential rate. Human populations, in which individuals live and reproduce for many years and in which reproduction is distributed throughout the year, grow exponentially.

Exponential population growth can be determined by dividing the change in population size (ΔN) by the time interval (Δt) for a certain population size (N),

$$\frac{\Delta N}{\Delta t} = rN.$$

The growth curve of these populations is smooth and becomes increasingly steeper over time. The steepness of the curve depends on the intrinsic rate of natural increase for the population. Human population growth has been exponential since the beginning of the 20th century. Much concern exists about the impact this growth will have, not only on the environment but on humans as well.

Insects and plants that live for a single year and reproduce once before dying are examples of organisms whose growth is geometric. In these species a population grows as a series of increasingly steep steps rather than as a smooth curve.

Logistic Population Growth

The geometric or exponential growth of all populations is eventually curtailed by food availability, competition for other resources, predation, disease, or some other ecological factor. If growth is limited by resources such as food, the exponential growth of the population begins to slow as competition for those resources increases. The growth of the population eventually slows nearly to zero as the population reaches the carrying capacity (K) for the environment. The result is an S-shaped curve of population growth known as the logistic curve. It is determined by the equation

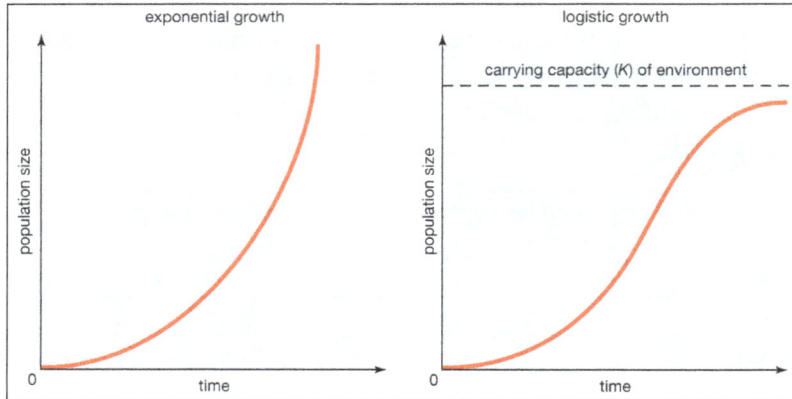

Carrying capacity; exponential versus logistic population growth
In an ideal environment (one that has no limiting factors) populations grow at an exponential rate.

The growth curve of these populations is smooth and becomes increasingly steep over time (left). However, for all populations, exponential growth is curtailed by factors such as limitations in food, competition for other resources, or disease. As competition increases and resources become increasingly scarce, populations reach the carrying capacity (K) of their environment, causing their growth rate to slow nearly to zero. This produces an S-shaped curve of population growth known as the logistic curve (right).

$$\frac{\Delta N}{\Delta t} = rN\left(\frac{K - N}{K}\right).$$

Population Fluctuation

As stated above, populations rarely grow smoothly up to the carrying capacity and then remain there. Instead, fluctuations in population numbers, abundance, or density from one time step to the next are the norm. Population cycles make up a special type of population fluctuation, and the growth curves in population cycles are marked by distinct amplitudes and periods that set them apart from other population fluctuations. In a few species, such as snowshoe hares (Lepus americanus), lemmings, Canadian lynx (Lynx canadensis), and Arctic foxes (Alopex lagopus), populations show regular cycles of increase and decrease spanning a number of years. The causes of these fluctuations are still under debate by population ecologists, and no single cause may provide an explanation for every species. Most major hypotheses link regular fluctuations in population size to factors that are dependent on the density of the population, such as the availability of food or the activities of specialized predators, whose numbers track the abundance of their prey through population high and low.

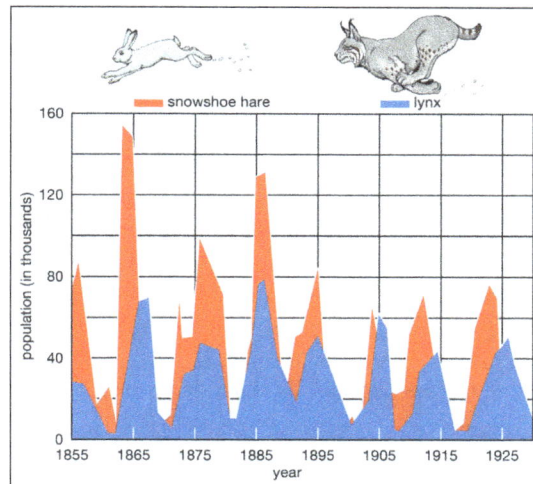

Cyclical fluctuations in the population density of the snowshoe hare and its effect
on the population of its predator, the lynx.

Factors affecting Population Fluctuation

Population ecologists commonly divide the factors that affect the size of populations into density-dependent and density-independent factors. Density-independent factors, such as weather and climate, exert their influences on population size regardless of the population's density. In contrast, the effects of density-dependent factors intensify as the population increases in size. For example, some diseases spread faster in populations where individuals live in close proximity with one another than in those whose individuals live farther apart. Similarly, competition for food and other resources rises with density and affects an increasing proportion of the population. The dynamics of most populations are influenced by both density-dependent and density-independent factors, and the relative effects of the factors vary among populations. Density-independent factors are known as limiting factors, while density-dependent factors are sometimes called regulating factors because of their potential for maintaining population density within a narrow range of values.

Population Cycles

Because many factors influence population size, erratic variations in number are more common than regular cycles of fluctuation. Some populations undergo unpredictable and dramatic increases in numbers, sometimes temporarily increasing by 10 or 100 times over a few years, only to follow with a similarly rapid crash. The populations of some forest insects, such as the gypsy moths (Lymantria dispar) that were introduced to North America, rise extremely fast. As with species that fluctuate more regularly, the causes behind such sudden population increases are not fully known and are unlikely to have a single explanation that applies to all species.

The size of other populations varies within tighter limits. Some fluctuate close to their carrying capacity; others fluctuate below this level, held in check by various ecological factors, including predators and parasites. The tremendous expansion of many populations of weeds and pests that have been released into new environments in which their enemies are absent suggests that predators, grazers, and parasites all contribute to maintaining the small sizes of many populations.

Invasive prickly pear cactus in Australia: Area in Queensland, Australia, covered with prickly pear cactus (Opuntia stricta), an invasive species that rapidly expanded its range after being introduced there.

Biological control of invasive prickly pear cactus in Australia: Area in Queensland, Australia, formerly covered with prickly pear cactus (Opuntia stricta). The cactus was introduced to the region, and three years later the moth borer (Cactoblastis cactorum) was introduced as a biological control agent to reduce populations of the cactus.

To control the explosive proliferation of these species, biological control programs have been instituted. With varying degrees of success, parasites or pathogens inimical to the foreign species have been introduced into the environment. The European rabbit (Oryctolagus cuniculus) was introduced into Australia in the 1800s, and its population grew unchecked, wreaking havoc on agricultural and pasture lands. The myxoma virus subsequently was released among the rabbit populations and greatly reduced them. Populations of the prickly pear cactus (Opuntia) in Australia and Africa grew unbounded until the moth borer (Cactoblastis cactorum) was introduced. However, many other similar attempts at biological control have failed, illustrating the difficulty in pinpointing the factors involved in population regulation.

Species Interactions and Population Growth

Interspecific Interactions

Community-level interactions are made up of the combined interactions between species within the biological community where the species coexist. The effects of one species upon another that derive from these interactions may take one of three forms: positive (+), negative (−), and neutral (0). Hence, interactions between any two species in any given biological community can take any of six forms:

1. Mutualism (+, +), in which both species benefit from the interaction.

2. Exploitation (+, −), in which one species benefits at the expense of the other.

3. Commensalism (+, o), in which one species benefits from the interaction while the other species neither benefits nor suffers.

4. Interspecific competition (−, −), in which both species incur a cost of the interaction between them.

5. Amensalism (−, o), in which one species suffers while the other incurs no measurable cost of the interaction.

6. Neutrality (o, o), in which both species neither benefit nor suffer from the interaction.

Lotka-volterra Equations

The effects of species interactions on the population dynamics of the species involved can be predicted by a pair of linked equations that were developed independently during the 1920s by American mathematician and physical scientist Alfred J. Lotka and Italian physicist Vito Volterra. Today the Lotka-Volterra equations are often used to assess the potential benefits or demise of one species involved in competition with another species,

$$dN_1 / dt = r_1 N_1 \left(1 - N_1 / K_1 - \alpha_{1,2} N_2 / K_2 \right)$$
$$dN_2 / dt = r_2 N_2 \left(1 - N_2 / K_2 - \alpha_{2,1} N_1 / K_1 \right)$$

Here r = rate of increase, N = population size, and K = carrying capacity of any given species. In the first equation, the change in population size of species 1 over a specific period of time (dN_1/dt) is determined by its own population dynamics in the absence of species 2 ($r_1 N_1[1 - N_1/K_1]$) as well as by its interaction with species 2 ($\alpha_{1,2}N_2/K_2$). As the formula implies, the effect of species 2 on species 1 ($\alpha_{1,2}$) in turn is determined by the population size and carrying capacity of species 2 (N_2 and K_2).

The possible outcomes of interactions between two species are predicted on the basis of the relative strengths of self-regulation versus the species interaction term. For instance, species 2 will drive species 1 to local extinction if the term $\alpha_{1,2}N_2/K_2$ exceeds the term $r_1 N_1(1 - N_1/K_1)$—though the term $\alpha_{1,2}N_2/K_2$ will exert a decreasing influence over the growth rate of species 1 as $\alpha_{1,2}N_2/K_2$ diminishes. Consequently, the first equation represents the amount by which the growth rate of species 1 over a specific time period will be reduced by its interaction with species 2. In the second equation, the obverse applies to the dynamics of species 2.

In the case of interspecific competition, if the effects of both species on each other are approximately equivalent with respect to the strength of self-regulation in each species, the populations of both species may stabilize; however, one species may gradually exclude the other over time. The competitive exclusion scenario is dependent on the initial population size of each species. For instance, when the interspecific effects of each species upon the abundance of its competitor are approximately equal, the species with the higher initial abundance is likely to drive the species with a lower initial abundance to exclusion.

The basic equations given above, describing the dynamics deriving from an interaction between two competitors, have undergone several modifications. Chief among these modifications is the development of a subset of Lotka-Volterra equations that calculate the effects of interacting predator and prey populations. In their simplest forms, these modified equations bear a strong resemblance to the equations above, which are used to assess competition between two species,

$$dN_{prey} / dt = r_{prey} \times N_{prey} \left(1 - N_{prey} / K_{prey} - \alpha_{prey, pred} \times N_{pred} / K_{pred}\right)$$

$$dN_{pred} / dt = r_{pred} \times N_{pred} \left(1 - N_{pred} / K_{pred} + \alpha_{pred, prey} \times N_{prey} / K_{prey}\right)$$

Here the terms N_{pred} and K_{pred} denote the size of the predator population and its carrying capacity. Similarly, the population size and carrying capacity of the prey species are denoted by the terms N_{prey} and K_{prey}, respectively. The coefficient $\alpha_{prey, pred}$ represents the reduction in the growth rate of prey species due to its interaction with the predator, whereas $\alpha_{pred, prey}$ represents the increase in growth rate of the predator population due to its interaction with prey population.

Several additional modifications to the Lotka-Volterra equations are possible, many of which have focused on the incorporation of influences of spatial refugia (predator-free areas) from predation on prey dynamics.

Metapopulations

Although the dynamics and evolution of a single closed population are governed by its life history, populations of many species are not completely isolated and are connected by the movement of individuals (immigration and emigration) among them. Consequently, the dynamics and evolution of many populations are determined by both the population's life history and the patterns of movement of individuals between populations. Regional groups of interconnected populations are called metapopulations. These metapopulations are, in turn, connected to one another over broader geographic ranges. The mapped distribution of the perennial herb *Clematis fremontii* variety *Riehlii* in Missouri shows the metapopulation structure for this plant over an area of 1,129 square km (436 square miles). There is, therefore, a hierarchy of population structure from local populations to metapopulations to broader geographic groups of populations and eventually up to the worldwide collection of populations that constitute a species.

As local populations within a metapopulation fluctuate in size, they become vulnerable to extinction during periods when their numbers are low. Extinction of local populations is common in some species, and the regional persistence of such species is dependent on the existence of a metapopulation. Hence, elimination of much of the metapopulation structure of some species can increase the chance of regional extinction of species.

The structure of metapopulations varies among species. In some species one population may be particularly stable over time and act as the source of recruits into other, less stable populations. For example, populations of the checkerspot butterfly (*Euphydryas editha*) in California have a metapopulation structure consisting of a number of small satellite populations that surround a large source population on which they rely for new recruits. The satellite populations are too small and fluctuate too much to maintain themselves indefinitely. Elimination of the source population from this metapopulation would probably result in the eventual extinction of the smaller satellite populations.

Edith's checkerspot butterfly (Euphydryas editha), male: This well-studied species inhabits a large portion of western North America and is known for being especially sensitive to annual variation in weather and climate.

In other species, metapopulations may have a shifting source. Any one local population may temporarily be the stable source population that provides recruits to the more unstable surrounding populations. As conditions change, the source population may become unstable, as when disease increases locally or the physical environment deteriorates. Meanwhile, conditions in another population that had previously been unstable might improve, allowing this population to provide recruits.

Overall, the population ecology and dynamics of all species is a complex result of their genetic structure, the life histories of the individuals, fluctuations in the carrying capacity of the environment, the relative influences of all the different kinds of density-dependent and density-independent factors that limit population growth, the spatial distribution of individuals, and the pattern of movement between populations that determines metapopulation structure. It is, therefore, not surprising that there are often great fluctuations in the numbers of individuals in local populations and that the long-term persistence of species may often require the conservation of many, rather than a few, populations.

Predator-Prey Model

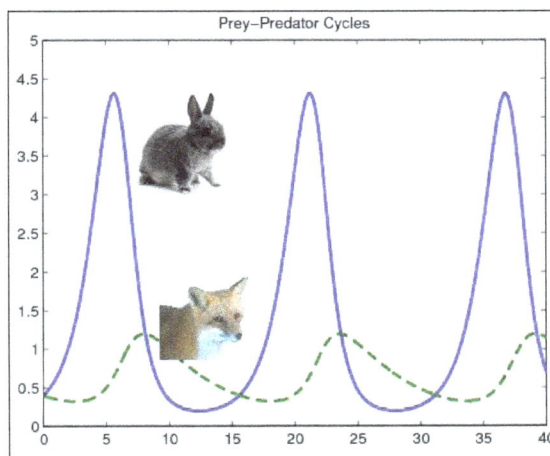

Figure: Periodic activity generated by the Predator-Prey model.

Predator-prey models are arguably the building blocks of the bio- and ecosystems as biomasses are grown out of their resource masses. Species compete, evolve and disperse simply for the purpose of seeking resources to sustain their struggle for their very existence. Depending on their specific settings of applications, they can take the forms of resource-consumer, plant-herbivore, parasite-host, tumours cells (virus)-immune system, susceptible-infectious interactions, etc. They deal with the general loss-win interactions and hence may have applications outside of ecosystems. When seemingly competitive interactions are carefully examined, they are often in fact some forms of predator-prey interaction in disguise.

A General Predator-prey Model

Consider two populations whose sizes at a reference time t are denoted by $x(t)$, $y(t)$, respectively. The functions x and y might denote population numbers or concentrations (number per area) or some other scaled measure of the populations sizes, but are taken to be continuous functions. Changes in population size with time are described by the time derivatives, $\dot{x} \equiv dx/dt$ and $\dot{y} \equiv dy/dt$, respectively, and a general model of interacting populations is written in terms of two autonomous differential equations,

$$\dot{x} = xf(x,y)$$

$$y \quad yg(x,y)$$

(i.e., the time t does not appear explicitly in the functions $xf(x,y)$ and $yg(x,y)$). The functions f and g denote the respective *per capita growth rates* of the two species. It is assumed that $df(x,y)/dy < 0$ and $dg(x,y)/dx > 0$. This general model is often called Kolmogorov's predator-prey model.

Lotka-volterra Model

The famous Italian mathematician Vito Volterra proposed a differential equation model to explain the observed increase in predator fish (and corresponding decrease in prey fish) in the Adriatic Sea during World War I. At the same time in the United States, the equations studied by Volterra were derived independently by Alfred Lotka to describe a hypothetical chemical reaction in which the chemical concentrations oscillate. The Lotka-Volterra model is the simplest model of predator-prey interactions. It is based on linear per capita growth rates, which are written as,

$$f = b - py$$

and $g = rx - d$.

- The parameter b is the growth rate of species x (the prey) in the absence of interaction with species y (the predators). Prey numbers are diminished by these interactions: The per capita growth rate decreases (here linearly) with increasing y, possibly becoming negative.

- The parameter p measures the impact of predation on $\dot{x}/x$.

- The parameter d is the death (or emigration) rate of species y in the absence of interaction with species x.

- The term rx denotes the net rate of growth (or immigration) of the predator population in response to the size of the prey population.

The Prey-Predator model with linear per capita growth rates is,

$$\dot{x} = (b - py)x$$

(Prey),

$$\dot{y} = (rx - d)y.$$

(Predators) This system is referred to as the Lotka-Volterra model: it represents one of the earliest models in mathematical ecology.

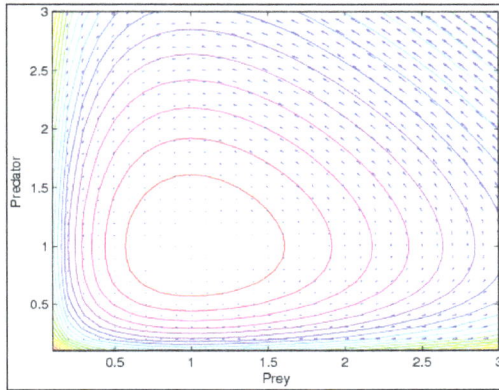

Figure: Prey-Predator dynamics as described by the level curves of a conserved quantity. The arrows describe the velocity and direction of solutions. In this simulation, the data are $d = r = b = d = 1$. There are equilibria at $x = 1, y = 1$ and at $x = 0, y = 0$.

The system can be integrated directly. In particular, any solution $(x(t),y(t))$ of the system satisfies the identity,

$$C = b \ln y(t) - py(t) - rx(t) + d \ln x(t)$$

For all t, where the constant $C = b \ln y(0) - py(0) - rx(0) + d \ln x(0)$ is determined by initial conditions and system parameters. (Here in x denotes the natural logarithm of x, etc.) The quantity on the right hand side of the identity above is referred to as a *conservation law*, since it is constant along any solution. Having a conserved quantity facilitates visualizing solutions. In the figure we draw level-set contours of the surface,

$$z = b \ln y - py - rx + d \ln x$$

in the first quadrant of the xy-plane. The contours describe solutions of the system determined by their initial data, and since they are closed curves, the solutions are periodic oscillations.

If $b > 0$, there are two equilibria, $x = 0, y = 0$ (extinction), and $x = d / r, y = b / p$ (co-existence), and the surface,

$$z = b \ln y - py - rx + d \ln x$$

has a single peak at the latter equilibrium. The contour lines in the figure describe the classic prey-predator cycles observed in ecological systems.

The model above has been derived independently in the following fields:

- Epidemics (b=0)

  ○ x are susceptible individuals and

  ○ y are infective individuals.

- Ecology

  ○ x are prey and

  ○ y are predators.

- Combustion theory

  ○ x and y are chemical radicals formed during $H_2 O_2$ combustion.

- Economics

  ○ x is the populace and

  ○ y is a predatory institution.

Kermack-McKendrick Model

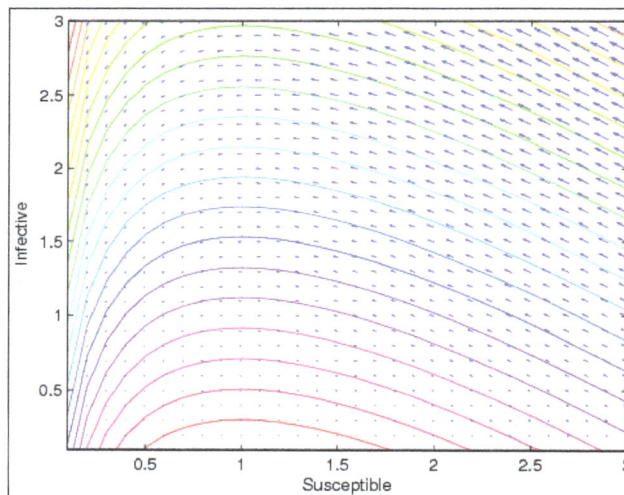

Kermack-McKendrick model of propagation of infectious disease.

There is herd immunity in predation and in epidemics. It is convenient to frame this in terms of epidemiology where now we refer to the prey as being susceptibles and the predators as being infectives. The infection dynamics is depicted by the graph $x \rightarrow y \rightarrow$, indicating that susceptibles can become infectives and that infectives can be removed from the process (e.g., through death, quarantine, or inoculation). Consider a time interval that is short compared to reproduction of the susceptible population, i.e., let $b = 0$. If the initial susceptible population is so large that $x(0) > d/r$, then we see from the second equation in the predator-prey model that initially $\dot{y} > 0$, which indicates that the infectives will initially more than replace themselves by passing on the infection. However, if this condition is not satisfied, the infective population will decrease. The critical value $R \equiv rx(0)/d = 1$ is referred to as being an epidemic threshold or a tipping point for the process.

The tipping point is at $x = 1.0$. If $x(0)$ is above this value, an epidemic will ensue. The severity can be estimated by tracing the curve emanating from $x(0)$ until it converges to the horizontal axis. This indicates the size of the susceptible population that avoids the infection. Note that the horizontal axis as drawn here begins at $x = 0.1$ to avoid the logarithmic singularity.

A short calculation shows that $x(t)$ converges to a constant, say $x(t) \to x*$, where $x*$ can be found by solving the equation $C = rx* - d \ln x*$, as shown in the figure. Surprisingly, this number is always greater than zero, which shows that some susceptibles will always survive.

This phenomenon, which is referred to as *herd immunity* is observed in practice; in fact, the number R is published regularly for various diseases and locales as an epidemic control measure. It reflects the fact that the susceptible population can be reduced to a level below which infectives will not increase. The model in this case is referred to as being the Kermack-McKendrick model of susceptible-infective interactions in epidemiology.

Jacob-Monod Model

Another approach to modeling the interaction between prey and predators was developed to account as well for organisms (such as bacteria) taking up nutrients. There is a limited uptake rate that such organisms are capable of, and the next model accounts for limited uptake rates. Suppose now that x denotes the size of a population of feeders and that they are feeding on a chemical species of concentration y. These are usually taken to represent concentrations of feeders and nutrients in solution rather than "head counts". The Jacob-Monod model is,

$$\dot{x} = \frac{Vy}{K+y}x$$

$$\dot{y} = -\frac{1}{V}\frac{Vy}{K+y}x$$

Where:

- V is the uptake velocity,

- K is the saturation constant, and

- Y is the yield of x per unit y taken up.

When $y = K$, the uptake velocity is $V/2$, half the maximum; in practice, $y = K$ is taken as a tipping point: If $y < K$, then uptake is ignored. This is the canonical model of nutrient uptake (nutrient y is taken up by species x), and it underlies many calculations in biology, microbiology, and food engineering (digestion, beer, etc.). This model was discovered independently in several diverse applications. It is akin to the Haldane-Briggs model and Michaelis-Menten model in biochemistry, the Jacob-Monod model in microbial ecology, and the Beverton-Holt model in fisheries. It serves as one of the important building blocks in studies of complex biochemical reactions and in ecology.

The quantity $C = x + Yy$ is conserved (since the derivative with respect to time of the right hand side is zero) where $C = x(0) + Yy(0)$.

Substituting this into the first equation gives,

$$\dot{x} = \frac{V(C-x)}{YK+(C-x)} x \, .$$

The solutions can be found by quadratures, and these show that if $x(0) > 0$, then $x(t) \rightarrow C$ as $t \rightarrow \infty$, at which point the nutrient has been depleted.

A typical use of this model is to describe a continuous-flow growth device, such as a chemostat, where there is continuous removal of nutrient and feeders and a continuous supply of fresh nutrient. The Jacob-Monod model is used to describe such a bacterial growth device, for example to determine conditions for a sustained dynamic equilibrium to exist by balancing growth due to uptake of nutrient with wash out of feeders.

Some predator-prey models use terms similar to those appearing in the Jacob-Monod model to describe the rate at which predators consume prey. More generally, any of the data in the Lotka-Volterra model can be taken to depend on prey density as appropriate for the system being studied. This is referred to as a functional response, an idea that is introduced. Several different forms of functional response have been used in population models, but the Jacob-Monod form, also called the Holling type 2 form by ecologists, is one of the more common ones. Many other investigations of predator-prey models have involved functional responses. For example, r is replaced by $r\ln(K/x), r(K-x)/(K+fx)$, (Smith, 1963), and $r((K/x)g-1)$ with $0 < g \le 1$,. Such mechanisms in the Lotka-Volterra model can stabilize or destabilize the system, for example resulting in predator extinction or in co-existence of prey and predators. This is in contrast to the plurality of cycles predicted by the original Lotka-Volterra model.

Logistic Equation

An interesting case for,

$$\dot{x} = \frac{V(C-x)}{YK+(C-x)} x$$

is when V and YK are very large compared to the other data in the model, but with their ratio being of moderate size, say $V/(YK) \approx r$.

Then we can ignore the second term in the denominator and get,

$$\dot{x} = r(C-x)x.$$

This is referred to as the *logistic equation*. It, too, has arisen in various disciplines, but one of its first appearance was due to Verhulst in the mid 19th century who used it to correct certain deviations of Malthus's model $(\dot{x}=rx)$ from certain human population data. The number C is now referred to as being the *carrying capacity* for the population; this corresponds to there being no remaining nutrient in the Jacob-Monod model.

The logistic equation can be solved in closed form by quadratures. This shows that,

$$x(t) \rightarrow C \text{ as } t \rightarrow \infty, \text{ if } x(0) > 0.$$

Predation with Time Delays: Chaos in Ricker's Reproduction Equation

Time delays occur in biological systems, and they can produce complicated dynamics. To model age structure (and other time delays) in a system, we take the approach that was introduced by Euler in the 18th century. Let $x(t)$ denote the number born at time t. Then $x(t-a)$ denotes the number who were born a units ago. Suppose that there is a nonlinear effect such that the number of newborns a units ago who can participate in reproduction at time t is $h(x(t-a))$ where the function h is called the *reproduction function*. This nonlinearity might be due to predation or environmental factors, Suppose those of age a have fertility $m(a)$. Then the equation below shows how many will be born at time t.

It is obtained by adding up across all ages those who can participate weighted by their fertility:

$$x(t) = x_0(t) + \int_0^t m(a)h(x(t-a))da,$$

Where $x_0(t)$ denotes the lingering contributions to later births of the initial population. In the case where $h(x) = x$ is a linear function, this equation is referred to as being the *renewal equation*, which is widely used in demography. Its solution can be found using Laplace transforms. However, if h is a nonlinear function, then that method of solution is not available.

The fertility function $m(a)$ describes the fertility of those of age a. If m is constant, then this equation is equivalent to a differential equation. The other extreme occurs when fertility is focused all at one age, as for salmon or cicadas. In this case, suppose that all fertility is due to those of age $a*$. We write $m(a) = r\delta(a - a*)$ where δ is Dirac's delta function and r is the reproduction rate.

The equation becomes,

$$x(t) = x0(t) + rh(x(t - a*)).$$

Let us focus only on the births at the times a*,2a*,3a*,......... ; we define $xn = x(na*)$ with $x_1 = x_0(a*)$. Substituting this in the equation above gives the recursion,

$$x_{n+1} = rh(x_n),$$

For n=1,2,3,4,... In the case of Malthus's model, $h(x) = x$, and the solution is a simple geometric progression $x_n = r^n x_0$. However, if h is as in Verhulst's case, where h(x) = x(C−x)$_+$ ≡ max (C −x,0), the recursion becomes,

$$x_{n+1} = rx_n(C - x_n)_+,$$

Whose iterates can exhibit quite wild behavior. This was first noticed by the ecologist W.E. Ricker used the function $h(x) = x\exp(-x)$ in studies of the dynamics of fisheries, although his work was largely ignored at the time. This reproduction function accounts for cannibalism (self predation) in that if the population is small, the model looks like Malthus's, but for large populations, reproduction is strongly suppressed. The work was later rediscovered by Robert May, who stimulated the now prominent area of chaos.

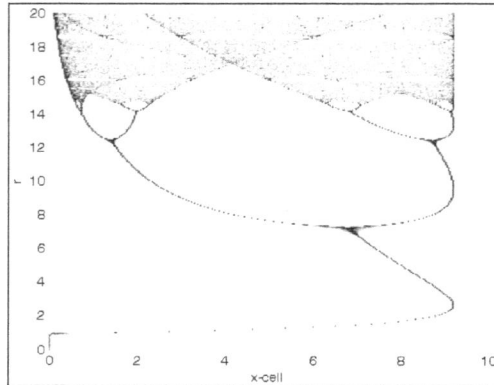

Dynamics of Ricker's population.

Chaotic dynamics can be illustrated by a simple computer simulation using Ricker's model. Since the maximum of h is 1, we need only consider values of $x \in [0, r]$. We divide this interval up into 300 equal bins. The parameter r is fixed at one of 300 equally spaced numbers in $r \in [0,20]$. The recursion is,

$$x_{n+1} = rx_n e^{-xn}.$$

Beginning with $x_1 = 1$, we determine the sequence $\{x50, x51, x52, x53, \ldots, x300\}$ and record in an array, say $D(r, x)$, the number of sequence elements that are in each bin. We only start counting after 50 initial iterates to avoid transients. The result is a surface $D(r, x)$ that describes the dynamics. We plot this record looking from above, as shown in the figure. The solution always settles into some structure, which might be highly complex. At first, there is a unique stable state. Near r = 8 this state bifurcates into an oscillation of period 2. Near $r=12$ the period 2 solution bifurcates into a period 4 solution, etc. By $r = 18$ the solution set is quite complicated. This is in the range of chaotic dynamics.

Allee Effect

Allee effects, named after the scientist W.C. Allee, are beneficial effects of conspecific interactions upon individuals. These benefits to reproduction, growth, and survival may generate behavioral effects, including preferences for living near other individuals. In addition, they have a number of population-level effects. Most importantly, they may affect the long-term persistence of a population if it cannot replace itself at a small size. In addition, Allee effects slow the rate of population spread and increase clumping over the landscape. The consequence of clumping is population subdivision, which can result in more complex population dynamics that improve persistence when compared to the dynamics of single isolated populations. All these processes have important consequences for the applied sciences of conservation biology, harvest management, and pest eradication.

Mechanisms that cause Allee Effects

A positive relationship between fitness and population size can be caused by a variety of mechanisms that affect reproduction and survival. A well established example, mate limitation, may result in undercrowding in species that reproduce sexually, because sexual reproduction requires contact between male and female gametes. Mate limitation reduces reproduction when plants or animals release gametes into the environment or when males and females have difficulty locating each other. When behaviors such as breeding, feeding, and defense are cooperative, they become more efficient or successful in larger social groups, resulting in increased reproductive success or survivorship. Although cooperative behaviors are most obvious in social vertebrates, such as prairie dogs, ungulates or birds, Allee effects resulting from group feeding or defense can also arise in insects, such as bark beetles, and aquatic organisms, such as cichlid fish.

Other mechanisms do not require cooperation in the behavioral sense, but merely the presence of conspecific individuals. For example, the per capita risk of predation is smaller in large prey populations than small prey populations. It is also known that the presence of multiple individuals can alter environmental or biotic conditions in favorable ways. Examples of such niche construction include reducing physical damage in intertidal zones or exclusion of competitors via allelopathy.

Finally, demographic and genetic mechanisms may give rise to Allee effects. In animals, active dispersal away from low-density populations can result in decreased rates of population growth. For many organisms, when population size is small, inbreeding depression can cause an Allee effect by reducing average fitness as population size declines. While these phenomena differ in form, in the way they affect fitness, and in which species are affected, they all result in the same general pattern: small populations suffer from reduced average individual fitness.

Evidence for Component Allee Effects

There is evidence from natural populations for component Allee effects due to all of these mechanisms. The most commonly observed mechanism is mate limitation, which causes Allee effects in both animals and plants (in the form of pollen limitation). Positive density dependence in survivorship due to either cooperative defense or predator satiation is also found across taxonomic groups. Indeed, because the tendency of predators to become satiated and stop increasing consumption depends on the predator rather than the prey, predator satiation has the potential to affect any population fed upon by a predator that does not numerically track the abundance of the small prey population, such as when a generalist predator feeds on multiple species. Evidence for the other mechanisms described above is less abundant, but each has been found in multiple taxonomic groups, and some studies have detected positive density dependence in reproduction or survival without identifying the mechanism, making it likely that there are other mechanisms that lead to component Allee effects. The population level impacts of these mechanisms are less clear.

| | Taxonomic Group | | | | | |
| | terr. arthr. | aq. invert | mamm. | bird | fish | plant |
Mechanism						
mate limitation	○	○	o	o	?	○
cooperative defense	o	o	○	○	○	?
predator satiation	?	○	○	o	○	?
cooperative breeding	?	?	○	○	o	?
cooperative feeding	○	?	?	o	?	?
dispersal	○	?	o	?	?	?
habitat amelioration	o	o	?	?	o	o
other/ unknown	o	?	?	o	o	○

The number of studies which attributed the
Allee effect in a species to a given mechanism, according to taxonomy.

Taxonomic groups for which there is evidence of Allee effects are terrestrial arthropods, aquatic invertebrates, mammals, birds, fish, and plants. Species were included if the mechanism was detected empirically in natural populations. Single studies can be represented by multiple evidence types and mechanisms. The area of the circle represents the relative number of species exhibiting each mechanism (smallest circle = 1 species, largest = 20 species).

Evidence for Demographic Allee Effects

Demographic Allee effects have been harder to demonstrate. Two reasons for this include:

1. Allee effects in one component of fitness may be offset at low density by increases in other components of fitness, such as decreased competition for resources.

2. Natural populations at low density are often difficult to detect and the high variance that results from small sample sizes obscures statistical analysis.

For these reasons, unusual cases of population growth are important to the ongoing study of Allee effects. In particular, monitoring data from invasive species provides an opportunity for improving understanding of Allee effects. Some likely effects of positive density dependence are easier to observe, extinction, for instance, but it is difficult to confirm that an Allee effect was to blame.

Dynamics of Populations with Allee Effects

The most dramatic consequences of Allee effects are associated with strong Allee effects, although weak Allee effects are also predicted to give rise to measurable dynamical differences. The difference between Allee effects and the classical (Malthusian) logistic theory is seen most clearly in the correlation between per capita growth rate and population size. This correlation can be generated using:

$$\frac{dx}{dt} = rx\left(1 - \frac{x}{k}\right)\left(\frac{x-a}{k}\right)$$

Where x is the population size a is the critical point, and k represents the force of competition and is known as the carrying capacity.

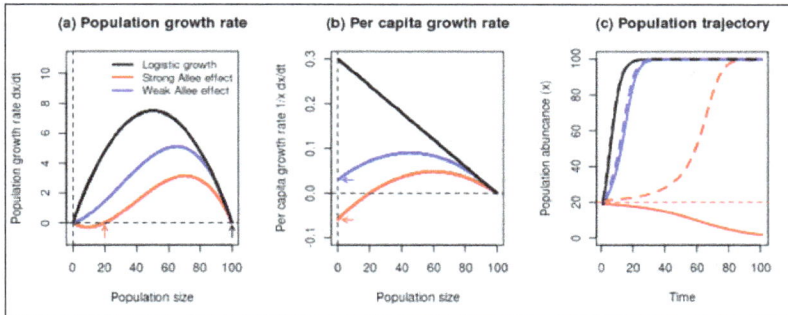

Figure: Allee effects are distinguished from other forms of density dependence by a positive association between average absolute individual fitness and population size over a finite interval.

When Allee effects are strong, a critical size occurs where the growth function dx/dt=f(x) intersects the horizontal line at zero (panel a, red arrow). This is an unstable equilibrium. Populations with abundance greater than this value will increase to carrying capacity (panel a, black arrow). Populations with abundance less than this value will decline to extinction. While carrying capacity is a stable equilibrium for both strong and weak Allee effects, and for logistic growth, extinction is stable only for strong Allee effects and unstable for weak Allee effects and logistic growth. Both kinds of Allee effects may be diagnosed by the positive association between per capita population growth rate and small population sizes, shown here to reach an intermediate maximum for weak Allee effects and for strong Allee effects (panel b). Strong and weak Allee effects may be distinguished by whether the y-intercept of the per capita growth rate is less than zero (indicating strong Allee effect, red arrow in panel b) or greater than zero (indicating weak Allee effect, blue arrow in panel b). Trajectories of populations with weak Allee effects are delayed compared with logistic growth, but qualitatively similar: populations initialized at any size will grow smoothly to carrying capacity (panel c). By contrast, the bistability of strong Allee effects means that eventual limit population size is determined by the initial condition (panel c). In panel c, the dashed lines for each population are for trajectories with initial size $N_0 = 21$ and the solid lines for trajectories with initial size $N_0 = 19$. From this plot, it is evident that the effects of small changes in initial population size are virtually indistinguishable for populations with logistic growth or weak Allee effects, but may be crucial for populations with strong Allee effects. Further, the delay between the trajectories for populations with weak Allee effects or logistic growth and the dashed red line for a population with a strong Allee effect initialized just greater than the critical population size at $N_0 = 21$, shows the very slow population takeoff caused by strong Allee effects, even when growth is positive.

Dividing both sides by the population size (x) yields the per capita population growth rate:

$$\frac{1}{x}\frac{dx}{dt} = r\left(1-\frac{x}{k}\right)\left(\frac{x-a}{k}\right).$$

The classical logistic theory predicts that per capita growth rate will not increase with population size (Figure b). By the definition above, a population with an Allee effect will exhibit an increase over some interval of population size. This distinction between Allee and non-Allee populations is the same regardless of whether the y-axis depicts per capita intrinsic rate of increase, the reproductive multiplier, λ, or lifetime reproductive output. Strong and weak Allee effects are distinguished according to whether the y-intercept falls below or above the replacement value (1/x dx/dt = 0 or λ = 1), respectively. These plots do not distinguish whether the Allee effect is endogenous to the dynamics of the population, such as result from mate limitation, or due to interactions with other species, such as the predator-induced Allee effect exhibited by the crown-of-thorns starfish *Acanthaster planci*.

Other dynamical consequences of strong Allee effects include:

1. A sigmoidal relationship between the probability of rapid extinction and the initial population size, a pattern which contrasts with the concave shape predicted by the classical logistic theory.

2. A critical area that the population must occupy for it to persist.

3. A bimodal stationary distribution of population size.

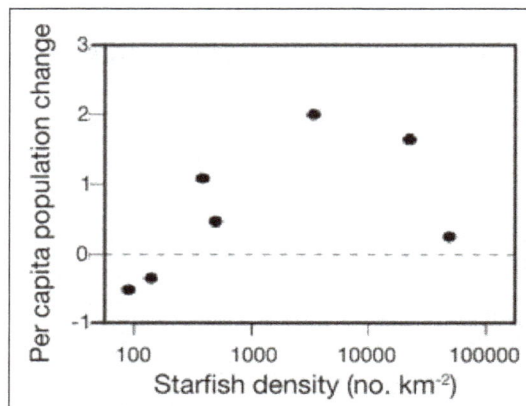

Figure: The relationship between per capita growth rate and starfish density show that the crown-of-thorns starfish (*Acanthaster planci*) is subject to a strong Allee effect.

In this case, the Allee effect is due to escape from fish predation at high density.

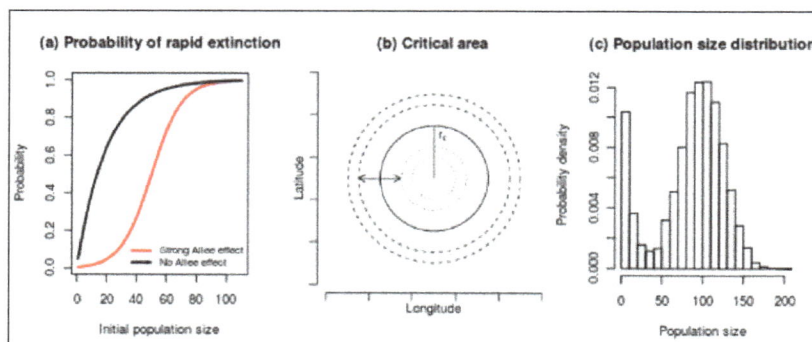

Figure: Three signatures of the Allee effect.

A sigmoidal probability of rapid extinction results from an inflection point corresponding to the critical population size (panel a). A critical area (panel b) results from the race between population growth at the center of a patch (where the population density exceeds the critical value) and dispersal at the periphery (where population density is below the replacement level). The critical area may be characterized by a radius rc that the population must exceed for the patch to persist. The solid line in panel b illustrates the critical area. The dashed (dotted) lines represent the areas subsequently occupied by populations that initially occupy the area just larger (smaller) than the critical area. The critical point in the growth dynamics also gives rise to a trough in the stationary distribution of population sizes once they have equilibrated (panel c). This trough causes the density to be bimodal with a lower mode in the vicinity of extinction and an upper mode at carrying capacity.

Evolution

The negative effects of low density, such as difficulty finding mates or increased vulnerability to predators, is expected to often result in strong selection for traits that reduce the influence of these mechanisms. For instance, rare species, which are typically at low density, often have adaptations that allow positive growth at low density, or they will become extinct. However, aggregation and sociality are not adaptations that eliminate Allee effects. Rather they increase population densities so that the densities at which Allee effects are manifest are avoided. Other traits may be the result of selection on low-density populations. For example, displays, calls, and pheromones all widen the area over which males and females perceive mates, thus reducing mate limitation. These adaptations can also increase fitness in other ways, however, such as signaling mate quality, so it is difficult to assign causality. Similarly, sperm storage, hermaphroditism, and parthenogenesis are all favored when chances of encountering mates are low, but may provide fitness benefits in other contexts as well, such as accelerated local adaptation. It is reasonable to conclude that species most likely to suffer Allee effects are those that usually have large populations but have suffered a recent reduction in size, such as due to habitat fragmentation or catastrophic natural events. Such populations will be more likely to suffer from mechanisms that reduce fitness at low density, especially if traits conferring fitness at low density incur a cost at high density. Finding such a trade-off may provide the best indication that a trait is an evolutionary response to persistence at low density.

Environmental Applications

When the size of populations subject to strong Allee effects is low, then these populations tend towards extinction. This fact argues for a thorough understanding of Allee effects and their mechanisms in order to develop sound management practices for a number of environmental issues. An obvious case is the conservation of rare species. It has been shown, for instance, that small patches of the flowering herb Clarkia concinna concinna attract pollinating insects in fewer numbers than do large patches. As a result, small patches are more prone to extinction. In general, persistence of species subject to strong Allee effects requires that a minimum population size — specific to each population — be distributed over a population-specific minimal area. This conclusion also applies to efforts to restore extirpated populations through captive breeding and reintroduction.

Another environmental issue that involves the Allee effect is the management of invasive species. Population biologists have long wondered why only a small fraction of introduced species ultimately persist in their introduced locations. A partial answer to this question is that many introductions are below the critical size associated with Allee effects. It follows that risk management for invasive species could be improved by a better understanding of the comparative biology of

Allee effects (i.e., the relative frequency and strength of Allee effects across species), coupled with a quantitative understanding of the rates at which introductions occur, a concept referred to as propagule pressure.

Even populations with weak Allee effects may require consideration of Allee effects for effective management. For instance, the invasive smooth cordgrass, Spartina alterniflora, is known to exhibit Allee effects due to pollination limitation. As a result, new patches of cordgrass initially grow quite slowly. An observer of such an invasion might wrongly conclude that this species' potential for spread is low. However, once the incipient colony grows to a certain size its potential to spread increases dramatically.

References

- Population-ecology, science: britannica.com, Retrieved 8 May, 2019

- Predator-prey-model: scholarpedia.org, Retrieved 28 January, 2019

- Allee-effect, earth-and-planetary-sciences: sciencedirect.com, Retrieved 19 July, 2019

- Allee-effects: nature.com, Retrieved 3 February, 2019

Chapter 5

Plant Ecology

The sub discipline of ecology which deals with the interactions between plants and other organisms is known as plant ecology. It also focuses on the distribution and abundance of plants and the effect which environmental factors have upon them. The topics elaborated in this chapter will help in gaining a better perspective about the different interactions which take place in the field of plant ecology.

Plant ecology is the study of the relationship of plants with the biotic (living organisms such as animals and other plants, bacteria, and fungi) and abiotic factors such as moisture, temperature, sunlight, soil (nutrients and salinity), and water (rainfall and water table) surrounding them. By the passage of time, the addressed issues regarding biosphere came into consideration. Though from the time of Alexander Von Humboldt (father of ecology) the known species of plants were about 20,000, now the number increased up to 40,0000 but the changes in elements of biosphere are increasing the issues such as loss of habitat, plant, animal, microflora, mutation, pollution, and soil sickness. Due to these issues, the most affected living organisms are plants which urged the scientists to investigate the root causes of such drastic changes and commotion to the plant ecology. According to the climate, human and animal interaction, flora, and fauna, the planet earth is categorized into biomes. There are about six major biomes with cutting clarity of subcategories. The largest biome is boreal/coniferous forest; however, the second largest biome is grasslands that are ubiquitous as compared to other biomes; tropical rainforest covers only 6% of the world, but they have the richest biodiversity; however, the hottest biome is desert with the minimal biodiversity; in contrast the coldest biome is alpine forest merely with considerable biodiversity. Specifically, the plant populations have dominantly occupied this globe; according to an estimation, 99.9% area of planet earth is covered with flora. About 350,000 species of plants excluding ferns, bryophytes, and algae are known and documented yet. Among them approximately 20% are under the risk of endangerment. The risk of endangerment or extinction due to natural and unnatural disasters has disturbed the whole food chain and web and is continually pushing toward the worst conditions.

Whenever ecological drift and loss in biodiversity of living organisms are discussed, it is generally apprehended that plants are vanishing due to overgrazing and animals are dying due to the inaccessibility of plants. But this whole globe is alive and functional on a single principle named balanced metabolic dynamics ratio between autotrophs (plants, producers) and heterotrophs (animals and microbes, consumers and decomposer). Basically, this trophic dynamics between producer and consumer within the biosphere is regulated by the transfer of energy from one part of the ecosystem to the other and even within the same ecosystem also known as energy flow in ecosystem. Except solar radiation (external source of energy), all the other energy systems are recycled and balance the dynamics of trophic level followed by complex metabolic mechanisms within the biosphere.

Drafting the origin of plant, their functional types and phylogenetic/evolutionary patterns are the most needed steps to timely track and record the drifts and risks to the ecosystem and biosphere.

As the current dynamics, composition and distribution of plants are altered thence, elaborating and redefining the relationship of plants with the factors encompassing them had led the flora and environment on the verge of endangerment are also expatiated, and many successful solutions to indemnify these issues are contributed by the scientists. Sustaining to this several concepts such as phylogeny, phenology, phytosociology, physiology, and anatomy of plants were used for modeling and surveying.

Terrestrial vegetation plays a phenomenal role in management of landscape and hydrological regime. Also the climatic change can be ameliorated by them as they could better regulate biogeological water cycle and sequestrate carbon cycle. The provision of protection against water resources by surface runoff leading toward flood attenuation, aquifer recharging, sea water leveling, water table leveling and fresh water management. Increase in temperature, variation in precipitation, and extreme events have potentially manifested the natural conversational and agricultural management regimes including an indirect risk that was constrained for social and human livelihoods.

Soil-Plant Interactions

Soil plays a key role in plant growth. Beneficial aspects to plants include providing physical support, water, heat, nutrients, and oxygen. Mineral nutrients from the soil can dissolve in water and then become available to plants. Although many aspects of soil are beneficial to plants, excessively high levels of trace metals (either naturally occurring or anthropogenically added) or applied herbicides can be toxic to some plants.

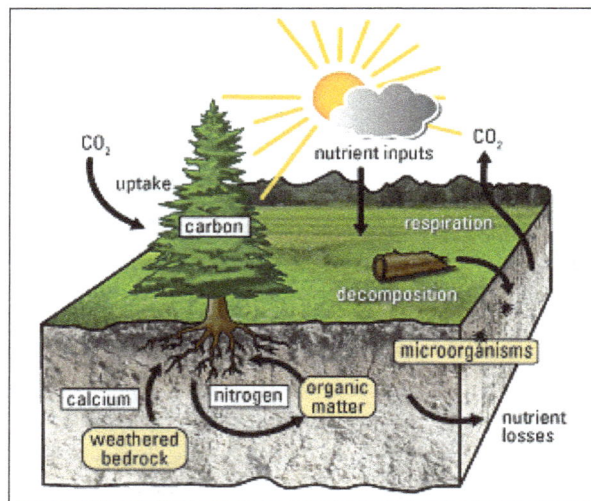

Soil-Plant Nutrient Cycle: This figure illustrates the uptake of nutrients by plants in the forest-soil ecosystem.

The ratio of solids/water/air in soil is also critically important to plants for proper oxygenation levels and water availability. Too much porosity with air space, such as in sandy or gravelly soils, can lead to less available water to plants, especially during dry seasons when the water table is low. Too much water, in poorly drained regions, can lead to anoxic conditions in the soil, which may be toxic to some plants.

Nutrient uptake by Plants

Several elements obtained from soil are considered essential for plant growth. Macronutrients, including C, H, O, N, P, K, Ca, Mg, and S, are needed by plants in significant quantities. C, H, and O are mainly obtained from the atmosphere or from rainwater. These three elements are the main components of most organic compounds, such as proteins, lipids, carbohydrates, and nucleic acids. The other six elements (N, P, K, Ca, Mg, and S) are obtained by plant roots from the soil and are variously used for protein synthesis, chlorophyll synthesis, energy transfer, cell division, enzyme reactions, and homeostasis (the process regulating the conditions within an organism).

Micronutrients are essential elements that are needed only in small quantities, but can still be limiting to plant growth since these nutrients are not so abundant in nature. Micronutrients include iron (Fe), manganese (Mn), boron (B), molybdenum (Mo), chlorine (Cl), zinc (Zn), and copper (Cu). There are some other elements that tend to aid plant growth but are not absolutely essential.

Micronutrients and macronutrients are desirable in particular concentrations and can be detrimental to plant growth when concentrations in soil solution are either too low (limiting) or too high (toxicity). Mineral nutrients are useful to plants only if they are in an extractable form in soil solutions, such as a dissolved ion rather than in solid mineral. Many nutrients move through the soil and into the root system as a result of concentration gradients, moving by diffusion from high to low concentrations. However, some nutrients are selectively absorbed by the root membranes, enabling concentrations to become higher inside the plant than in the soil.

Plant-Plant Interactions

In plant communities each plant might interact in a positive, negative, or neutral manner. Plants often directly or indirectly alter the availability of resources and the physical habitat around them. Trees cast shade, moderate temperature and humidity, alter penetration of rain, aerate soil, and modify soil texture. Plant neighbors may buffer one another from stressful conditions, such as strong wind. Some plants make contributions to others even after they die. Trees in old-growth forests that fall and decompose ("nurse" logs) make ideal habitat for seeds to sprout, and such a log may be covered with thousands of seedlings. While effects on the physical habitat are consistent aspects of communities, plant-to-plant competition to preempt resources also takes place, and in some instances chemical interactions occur between species.

Commensalism occurs as one species lives in a direct association with another (the host), gaining shelter or some other environment requisite for survival and not causing harm or benefit to the host. Orchids and bromeliads (Neoregelia spp.) live on the trunk or branches of their host, gaining water and nutrients from the air or bark surface without penetrating host tissue. Stocky roots and xeromorphic leaves that help gain and retain water are characteristic of vascular epiphytes (epiphyte means to live upon another). Bryophyte, lichen, and fern epiphytes are so abundant in the tropical rain forest that they often embody more plant material than their host trees. Another facilitation is illustrated by seedling growth of the Saguaro cactus

(Cereus giganteus), which typically occurs in the shade of paloverde trees or other plants, which create a better water-relationship environment for the cactus and protect it from the negative effects of the intense sun. Farming practices often use "nurse" plants to create a temporary improvement in the environment for the main crop. For example, oat and alfalfa may be seeded together so that oat shades and maintains better soil surface moisture for the emerging alfalfa seedlings.

Direct plant-plant contacts that benefit both organisms are termed mutualism. Taking the broader view of plants to include microorganisms, a good example of this arrangement is the association of legumes and nitrogen-fixing bacteria that live within legume root nodules. The legume benefits by obtaining nitrogen from the bacteria, while the bacteria gain necessary carbohydrate energy from legume photosynthesis. The free-living bacteria actually change and become bacteroids, no longer able to live outside the roots. The vast majority of higher plants have fungal-root associations called mycorrhizae. The vascular plants benefit because the fungus is much better at absorbing and concentrating phosphorus (and perhaps other mineral nutrients) than the root tissue, while the fungus gains a source of carbon compounds from the plant.

Parasitic plant-plant interactions are harmful to the host. A number of plants (e.g., dodders, broomrakes, and pinedrops) do not contain chlorophyll and cannot photosynthesize. They parasitize green plants by penetrating the outer tissue of the host plant with haustoria (rootlike projections), which eventually tap the water and food-conducting tissue. Mistletoe also form haustoria but the primary function of these structures is obtaining water, as this partially parasitic plant is capable of manufacturing its own food by photosynthesis. Witchweed (Striga spp.) has green leaves but is an obligate parasitic weed that causes tremendous crop losses to tropical-origin cereal grain crops and legumes. Witchweed has evolved so that chemicals from the host plant have become signals for witchweed seed to germinate and attach to the host. Subsequently, witchweed penetrates the host roots and steals water, minerals, and hormones. Strangler fig is a tree that germinates high in the host tree and sends roots to the ground, eventually killing the host when the fig roots and vines surround and strangle the flow of sugars in the host.

It is rare that plants are unaffected by neighboring plants. Negative effects on one of the neighbors are referred to as interference, and they include competition and allelopathy. Competition, the situation in which one plant depletes the resources of the environment required for growth and reproduction of the other plant, is the most common plant-plant phenomenon in nature. Members of plant associations that are more successful at gaining major resources—water, nutrients, light, and space—have the advantage and typically dominate the community. Competitive advantage may result from a plant's season of growth, growth habit, or morphological features such as depth of rooting, and special physiological capabilities like differences in rate of photosynthesis. In contrast to competition, allelopathic interference is the result of a plant adding toxic chemicals to the environment that inhibit the growth and reproduction of associated species or those that may later grow in the area. Many negative effects on target species probably occur from a combination of competition and allelopathy. Chemicals released from one plant may also be a communication to other plants, causing germination (e.g., Striga) or signaling defense responses to insect attack.

Plant-Animal Interactions

The ways in which certain animals and plants interact have evolved in some cases to make them interdependent for nutrition, respiration, reproduction, or other aspects of survival.

Ecology represents the organized body of knowledge that deals with the relationships between living organisms and their nonliving environments. Increasingly, the realm of ecology involves a systematic analysis of plant-animal interactions through the considerations of nutrient flow in food chains and food webs, exchange of such important gases as oxygen and carbon dioxide between plants and animals, and strategies of mutual survival between plant and animal species through the processes of pollination and seed dispersal.

A major example of animal-plant interactions involves the continual processes of photosynthesis and cellular respiration. Green plants are classified as ecological producers, having the unique ability, by photosynthesis, to take carbon dioxide and incorporate it into organic molecules.

Taking the products of photosynthesis and chemically breaking them down at the cellular level to produce energy for life activities. Carbon dioxide is a waste product of this process.

Mutualism

Mutualism is an ecological interaction in which two different species of organisms beneficially reside together in close association, usually revolving around nutritional needs.

One such example is a small aquatic flatworm that absorbs microscopic green algae into its tissues. The benefit to the animal is one of added food supply. The mutual adaptation is so complete that the flatworm does not actively feed as an adult.

The algae, in turn, receive adequate supplies of nitrogen and carbon dioxide and are literally transported throughout tidal flats in marine habitats as the flatworm migrates, thus exposing the algae to increased sunlight. This type of mutualism, which verges on parasitism, is called symbiosis.

Coevolution

Coevolution is an evolutionary process wherein two organisms interact so closely that they evolve together in response to shared or antagonistic selection pressure. A classic example of coevolution involves the yucca plant and a species of small, white moth (Tegitecula).

The female moth collects pollen grains from the stamen of one flower on the plant and transports these pollen loads to the pistil of another flower, thereby ensuring cross-pollination and fertilization. During this process, the moth will lay her own fertilized eggs in the flowers' undeveloped seed pods.

The developing moth larvae have a secure residence for growth and a steady food supply. These larvae will rarely consume all the developing seeds; thus, both species (plant and animal) benefit.

Although this example represents a mutually positive relationship between plants and animals, other interactions are more antagonistic. Predator-prey relationships between plants and animals are common. Insects and larger herbivores consume large amounts of plant material. In response to this selection pressure, many plants have evolved secondary metabolites that make their tissues unpalatable, distasteful, or even poisonous. In response, herbivores have evolved ways to neutralize these plant defenses.

Mimicry and Nonsymbiotic Mutualism

Mimicry: In mimicry, an animal or plant has evolved structures
or behavior patterns that allow it tomimic either its surroundings or another
organism as a defensive or offensive strategy.

Certain types of insects, such as the leaf hopper, walking stick, praying mantis, and katydid (a type of grasshopper), often duplicate plant structures in environments ranging from tropical rain forests to northern coniferous forests. Mimicry of their plant hosts affords these insects protection from their own predators as well as camouflage that enables them to capture their own prey readily.

Certain species of ambush bugs and crab spiders have evolved coloration patterns that allow them to hide within flower heads of such common plants as goldenrod, enabling them to ambush the insects that visit these flowers.

In nonsymbiotic mutualism, plants and animals coevolve morphological structures and behavior patterns by which they benefit each other but without living physically together.

This type of mutualism can be demonstrated in the often unusual shapes, patterns, and colorations that more advanced flowering plants have developed to attract various insects, birds, and mammals for pollination and seed dispersal purposes. Accessory structures, called fruits, form around seeds and are usually tasty and brightly marked to attract animals for seed dispersal.

Although the fruits themselves become biological bribes for animals to consume, often the seeds within these fruits are not easily digested and thus pass through the animals' digestive tracts unharmed, sometimes great distances from the parent plant. Some seeds must pass through the digestive plant of an animal to stimulate germination.

Other types of seed dispersal mechanisms involve the evolution of hooks, barbs, and sticky substances on seeds that enable them to be easily transported by animal fur, feet, feathers, or beaks. Such strategies of dispersal reduce competition between the parent plant and its offspring.

Pollinators

Pollinators.

Because structural specialization increases the possibility that a flower's pollen will be transferred to a plant of the same species, many plants have evolved a vast array of scents, colors and nutritional products to attract pollinators. Not only does pollen include the plant's sperm cells; it also represents a food reward.

Another source of animal nutrition is a substance called nectar, a sugar-rich fluid produced in specialized structures called nectarines within the flower or on adjacent stems and leaves.

Assorted waxes and oils are also produced by plants to ensure plant-animal interactions. As species of bees, flies, wasps, butterflies, and hawk-moths are attracted to flower heads for these nutritional rewards, they unwittingly become agents of pollination by transferring pollen from stamens to pistils.

Some flowers have evolved distinctive, unpleasant odors reminiscent of rotting flesh or feces, thereby attracting carrion beetles and flesh flies in search of places to reproduce and deposit their own fertilized eggs.

As these animals copulate, they often become agents of pollination for the plant itself. Some tropical plants, such as orchids, even mimic a female bee, wasp, or beetle, so that the insect's male counterpart will attempt to mate with them, thereby encouraging precise pollination.

Among birds, hummingbirds are the best examples of plant pollinators. Various types of flowers with bright, red colors, tubular shapes, and strong, sweet odors have evolved in tropical and temperate regions to take advantage of hummingbirds' long beaks and tongues as an aid to pollination.

Because most mammals, such as small rodents and bats, do not detect colors as well as bees and butterflies do, some flowers instead focus upon the production of strong, fermenting, or fruit-like odors and abundant pollen rich in protein. In certain environments, bats and mice that are primarily nocturnal have replaced day-flying insects and birds as pollinators.

Plant Resistance against Herbivory

Interactions between plants and insect herbivores are important determinants of plant productivity in managed and natural vegetation. In response to attack, plants have evolved a range of defenses to reduce the threat of injury and loss of productivity. Plants are exposed to threats of resource loss by herbivory in natural conditions experiencing damage; to mitigate losses many plant species develop defensive traits against herbivores, such as primary and secondary metabolites. Among herbivores are many arthropods, mollusks, vertebrates, and nematodes, and these groups consume between 5 and 20% of plant biomass annually.

The cost on investing in defense can be quantified in reduced growth, lower photosynthetic production, and reduced plant fitness. Plant defenses reduce the ability of herbivores to obtain nutrients from plant tissue. Plants with diminished defense capability may suffer greater herbivore damage and exhibit lower overall fitness under conditions of herbivore stress than well-defended plants.

Plants respond to herbivory through various morphological, biochemical, and molecular mechanisms and exhibit multifactorial traits against herbivory that are constitutively expressed or induced upon attack. The plant defense activated upon herbivory is a complex network of different pathways composed of direct and indirect defenses. Direct defense compounds such as glucosinolates or protease inhibitors directly influence the insect performance and feeding behavior, while indirect defenses like emission of volatile organic compounds after herbivore attack function as attractant for parasitic wasp which in turn predate on the attacker. While plants develop new defense compounds or mechanisms to enhance the resistance against herbivores, their attackers find new ways to bypass or detoxify these.

Insect herbivory induces several internal signals from wounded tissues, including calcium ion fluxes, phosphorylation cascades, and systemic and jasmonate signaling. These are perceived in undamaged tissues, which thereafter reinforce their defense producing low molecular weight defense compounds. Some compounds produced by plants constitutively or induced by herbivore damage are toxic or impair gut function in arthropod; examples include alkaloids, benzoxazinoids, glucosinolates, and terpenoids.

Added to this, there are some other defense mechanisms, such as mechanical defenses, indirect defenses, interactions with other organisms, etc. In this review, we focus in different traits defensive in plants and its effect on population dynamics and evolution in both plants and invertebrates.

Finally, we integrate all traits in a specific example in Pinus genera. Interactions between plants and insect herbivores are important determinants of plant productivity in managed and natural vegetation. In response to attack, plants have evolved a range of defenses to reduce the threat of injury and loss of productivity. Plants are exposed to threats of resource loss by herbivory in natural conditions experiencing damage; to mitigate losses many plant species develop defensive traits against herbivores, such as primary and secondary metabolites. Among herbivores are many arthropods, mollusks, vertebrates, and nematodes, and these groups consume between 5 and 20% of plant biomass annually.

The cost on investing in defense can be quantified in reduced growth, lower photosynthetic production, and reduced plant fitness. Plant defenses reduce the ability of herbivores to obtain nutrients from plant tissue. Plants with diminished defense capability may suffer greater herbivore damage and exhibit lower overall fitness under conditions of herbivore stress than well-defended plants.

Plants respond to herbivory through various morphological, biochemical, and molecular mechanisms and exhibit multifactorial traits against herbivory that are constitutively expressed or induced upon attack. The plant defense activated upon herbivory is a complex network of different pathways composed of direct and indirect defenses. Direct defense compounds such as glucosinolates or protease inhibitors directly influence the insect performance and feeding behavior, while indirect defenses like emission of volatile organic compounds after herbivore attack function as attractant for parasitic wasp which in turn predate on the attacker. While plants develop new defense compounds or mechanisms to enhance the resistance against herbivores, their attackers find new ways to bypass or detoxify these.

Insect herbivory induces several internal signals from wounded tissues, including calcium ion fluxes, phosphorylation cascades, and systemic and jasmonate signaling. These are perceived in undamaged tissues, which thereafter reinforce their defense producing low molecular weight defense compounds. Some compounds produced by plants constitutively or induced by herbivore damage are toxic or impair gut function in arthropod; examples include alkaloids, benzoxazinoids, glucosinolates, and terpenoids.

Added to this, there are some other defense mechanisms, such as mechanical defenses, indirect defenses, interactions with other organisms, etc. In this review, we focus in different traits defensive in plants and its effect on population dynamics and evolution in both plants and invertebrates. Finally, we integrate all traits in a specific example in Pinus genera.

Induced Defences

Plants respond to herbivore attack through a dynamic defense system that includes structural barriers, toxic chemicals, and attraction of natural enemies of target pests. Both defense mechanisms may be present constitutively or induced after damage by the herbivores. Most of chemicals are produced in response to herbivore attack. Induced defenses make the plants phenotypically plastic, and high variability in defensive chemical exhibits a better defense.

The induced defenses occur when past or current herbivory is a reliable cue of future attack and defenses are costly; while in environments where herbivory is constantly high, constitutive defenses should be favored.

Herbivorous insects produce oral secretions which contain compounds that elicit plant responses and plant elicitor peptides prevalence across wide-ranging plant families. In response, plant produces diverse chemical active compounds such as benzyl cyanide, fatty acid-amino acid conjugates, and proteins such as β-glucosidase. Plants can recognize herbivore elicitor and initiate a cascade of responses, including changes in plasma membrane potential and activation of networks of kinases and phytohormones. Three major plant hormones, jasmonic acid (JA), salicylic acid (SA), and ethylene (ET), function in a complex regulatory network essential in herbivore-induced defense responses.

Chemical Compounds in Plant Defense

Plants produce defensive metabolites, which do not affect the normal vegetative growth and development, but reduce the palatability of tissues in which are produced. Can be constitutive stored as inactive forms or induced in response to insect or microbe attack. The defensive metabolites are bioactive specialized compounds used to protect plant against herbivores, and these compounds can use as target systems unique to herbivores, such as the nervous, digestive, and endocrine organs, may act as repellents for generalist herbivores, while specialists are forced to invest resources in detoxification mechanisms.

Plant defense include changes in transmembrane potential immediately upon herbivory damage and are tightly followed by changes in the intracellular Ca^{2+} concentration and generation of H_2O_2. Kinases phytohormone jasmonic acid (JA), ethylene (ET), salicylic acid (SA), and nitric oxide (NO) are detectable within minutes. After roughly 1 h, gene activation is followed by metabolic changes.

Antinutritive proteinase inhibitors (PINs) are locally and systemically induced upon insect attack, but many other proteins contribute to antiherbivory responses. Enzymes such as polyphenol oxidase a threonine deaminase limit protein availability in the midgut, whereas others destabilize insect peritrophic membranes. Plants also draw upon a complex arsenal of small-molecule chemical defenses including terpenoids, alkaloids, phenylpropanoids, glucosinolates, lipids, and non-protein amino acids.

Volatiles which can alert neighbor plants or tissues to potential attacks are promoted by herbivory and are a complex blend. Volatiles induces indirect defenses inhibits oviposition and attracts natural enemies such as parasitoids and predators.

Alkaloids

Efficient feeding deterrents against herbivore group of compounds are the alkaloids, particularly such derived from quinolizidine, like cytisine and sparteine. These molecules are alkaline and contain nitrogen in a heterocyclic ring. Alkaloids are biosynthesized in roots from amino acids and probably are involved in defense against insect herbivory. Twenty percent of vascular plants synthesized alkaloids, particularly in plant families Leguminosae, Liliaceae, Solanaceae, and Amaryllidaceae.

Phenolics

Phenolics are produced by plants as compounds able to repel herbivores, inhibit enzymes, attract pollinators and fruit dispersers, absorb UV radiation, and decrease competition between plant neighbors. There are approximately 10,000 plant phenolics derived from shikimic y/o malonic acids. Phenolics can bind covalently to herbivore's digestive enzymes and inactivate them or halt the growth and development of larvae. Phenolics can be regulated for external conditions like light and nutrients; when a plant is stressed, it produces less phenolics than nonstressed plants.

Terpenoids

The most diverse class of bioactive natural products in plants is terpenoids, with approximately 40,000 structures. Terpenoids are synthesized from acetyl-CoA and play a role in plant defense, can act like active compounds in resin or as volatiles, repellents, and toxins, or can modify development in herbivores. Another characteristic in monoterpenes and sesquiterpenes is its ability to form essential oils, like limonene in citrus plants; these essential oils have repellent and toxic effects on insects. Many terpenoids can have synergistic effects upon release.

Nonprotein Amino Acids

Amino acid g-aminobutyric acid (GABA), a four carbon nonproteinogenic widespread in animals, plants, and microorganisms, can be implicated in defense responses. Wounding plant tissue and cell disruption caused by feeding insects is sufficient to induce rapid jasmonate-independent GABA synthesis and accumulation. When ingested the elevated GABA levels become toxic for the insects. GABA is synthesized by decarboxylation of L-glutamate bay glutamate decarboxylases (GAD) in shoots and roots and is a component in a plant's first line of general, rapid defense against invertebrate pests.

One metabolite induced in plants is tyrosine, which can be redirected into other primary and secondary metabolites, and its accumulation in excess in young leaves may not be adaptive as they would persist once the leaf is full in size and protected by toughness. In contrast to tyrosine, physiological constraints on catabolism may be selected against induction of phenolics and saponins. When plants exceed the capacity to store constitutive secondary metabolites could avoid autotoxicity.

Sulfur

Sulfur is a crucial element for plants, determining plant development, maintenance, and resistance to environmental stress. Sulfur is taken up by plants as inorganic sulfate and incorporated in different sulfated metabolites including glucosinolates, selected flavonoids, phytosulfokines, and hormones by distinct pathways. Some sulfated metabolites function in plant defense against pathogens and herbivores such as defensin and thionin peptide, antimicrobial defenses with widespread distribution, whereas antifeedant glucosinolates are limited to the Brassicales order. Bacillus subtilis activates plant growth by producing IAA y/o gibberellins and emits volatile metabolites (VOCs), which can activate transcripts related to cell wall modifications, primary and secondary metabolism, stress responses, hormone regulation, iron homeostasis, and sulfur-rich aliphatic and indolic glucosinolates. Plants exposed to Bacillus subtilis with elevated glucosinolates exhibit greater protection against generalist herbivores. Then, plant-growth-promoting rhizobacteria can enhance plant sulfur assimilation and integrate in plant defense.

Lipids

Fatty acids (FAs) are essential macromolecules present in all living organisms, are the major source of reserve energy, are essential components of cellular membranes, and are implicates as signaling molecules, modulating normal and disease-related physiologies in microbes, insects, animals, and plants. In plants, fatty acids regulate salt, drought, heavy metal tolerance, and herbivore feeding especially by JA is a FA derivate molecule. In Nicotiana attenuata fatty acid-amino acid conjugates (FACs) in the herbivore Manduca sexta oral secretions are the major elicitors that induce herbivory-specific signaling. FAs increased plant defense against pathogens and insects by stimulation of key short- and long-term regulatory process.

Simulated herbivory dramatically increased salicylic acid-induced protein kinase (SIPK) activity and jasmonic acid (JA) levels in damaged leaves and undamaged systemic leaves, whereas wounding alone had no detectable systemic effects. The activation of SIPK and elevation of JA in specific systemic leaves increase in the activity of an important antiherbivore defense, trypsin proteinase inhibitor (TPI). Then, N. attenuata can identify FACs produced by herbivory in damaged leaves and activate MAPK and JA signaling for activated defenses.

Another lipids produced by plants are alkamides. Natural alkamides are often insecticidal. Chrysanthemum cultivars show a wide variation in degree of host-plant resistance to the western flower thrips Frankliniella occidentalis. Extracts of chrysanthemum leaves revealed the presence of an unsaturated isobutylamide, N-isobutyl-(E,E,E,Z)-2,4,10,12-tetradecatetraen-8-ynamide. Alkamides account for natural host resistance to thrips. The participation of alkamides in host resistance to insects can be due to their role as elicitors of plant defense responses. For instance, it has been reported that linolenoyl-L-glutamine, an amide produced in oral secretions of caterpillars, is able to induce the production of volatile chemicals from plants that attract predators and parasites of the caterpillar while it feeds.

Jasmonic Acid and Ethylene

Jasmonic acid (JA) is an important regulator of defense responses against chewing insects, necrotrophic pathogens, and cell-content feeders such as spider mites and thrips. Herbivores stimulate JA production by octadecanoid pathway. In Arabidopsis, JA is conjugated with isoleucine through the enzyme jasmonoyl isoleucine conjugate synthase1 (JAR1) that conjugates binding to the F-box protein coronatine insensitive1 (COI1) and degrades jasmonate ZIM domain (JAZ) repressor proteins. Then, JA-responsive genes, including JAZ, which involves a negative feedback loop are activated. There are two possible pathways: MYC2 regulates positively vegetative storage protein 2 (VSP2) and lipoxygenase 2 (LOX2), which are JA-responsive inducible by wound. The another pathway implicates the ethylene response factor (ERF) (JA and ET are synergic) and induces ERF1 and ORA59; both are JA/ET-responsive transcription factors which regulate responsive genes like plant defensin 1.2 (PDR1.2). MYC2 regulates defense against herbivores, and ERF is involved in induced defense especially against necrotrophic pathogens.

Salicylic Acid

Salicylic acid (SA) is an essential signaling molecule that mediates pathogen-triggered signals perceived by different immune receptors to induce downstream defense responses. SA is a small phenolic phytohormone, which plays a major role in mediating defense; its accumulation is essential for induction of defense responses.

Induced plant responses are regulated by SA when herbivores bite phloem. Plant responses synthesizing SA from chorismate by isochorismate and phenylalanine ammonium lyase pathways. Increases in SA concentrations lead to nuclear translocation of pathogenesis-related genes 1 (NPR1), which results in the expression of defense proteins, the pathogenesis-related (PR) proteins.

When a plant faces multiple herbivore attack, induced defense is regulated through interconnection of the JA, SA, and ET signal transduction pathways. Cross talk between JA and SA signaling is mutually antagonistic, resulting in the prioritization of SA-dependent defense responses over JA-dependent responses or vice versa.

Mechanical Defenses

The first layer of defense in plant is mechanical, and the major components contributing to mechanical defenses are trichomes. These structures negatively influence on herbivore feeding behavior and insect mobility. Another trait in plant defense is the palatability, and one form to modify this character is to produce dense trichomes; for example, in Phaedon species, the host preference of adult beetles was less for Brassica cultivars that produced dense trichomes, while adult beetles were inclined to attack glabrous leaves. That is particularly important on young leaves of hairy plants, which produce denser trichomes than those of mature leaves. Therefore, trichomes might play an important role in the defense of younger leaves and contribute to future development of leaves. Trichomes tend to be more effective against insects that are small relative to trichome size; additionally, trichomes tend to deter sap-feeding or leaf-chewing insects to a greater extent than those feeding within plant tissues. Spinescence, including spines, thorns, and prickles, also defends the plants against many insects.

Epicuticular waxes form a slippery film or crystals that prevent from attaching to the plant surface, oviposition, or feeding. The biosynthesis and composition of waxes vary during plant development, and the physical-chemical properties of the cuticle respond on changes in season and temperature.

Another mechanical defense is to deposit granular minerals in tissues that deter insect attack and feeding. For example, Si accumulation, especially in Poaceae family, which is abrasive, damages herbivore feeding structures and reduces digestibility. Si accumulation can be induced by herbivory. Si in leaf surface can be abrasive in grasses with silicified spines, while others deposited Si in short cells. Si allocation to spines impacts palatability, while allocation to short cells may impact digestibility.

The cell walls of leaves are also reinforced during the feeding through the use of different macromolecules, such as lignin, cellulose, suberin, and callose, together with small organic molecules, such as phenolics and Si.

Good few plants contain laticifers and resin ducts that canals produce and store latex and resins under internal pressure; when the channels are broken, they are secreted and might entrap or intoxicate the herbivore. However, several specialist herbivores can block the flow of latex cutting the leaf veins, for example, the milkweed beetles Labidomera clivicolis, Tetraopes melanurus, and T. tetrophthalmus for feeding Asclepias cut veins and wait stop flow.

Oleoresins produced by conifers are a blend of terpenoids and phenolics accumulated in intercellular channels. When bark beetles bite that channels resin flow and get out the insect until outside, when oleoresins solidifying.

Indirect Defenses

Indirect defense can be used when plants attract, nourish, or house other organisms to reduce enemy pressure. For example, ant association in Mallotus japonicus (Euphorbiaceae) the damage leaf areas of ant excluded plants were much larger than those of control plants in middle-age leaves. This is done by producing volatiles, extrafloral nectar, food bodies, and nesting or refuge sites.

Extrafloral nectar is secreted on leaves and shoots to attract predators and parasitoids and consists mainly of sugars, amino acids, lipids, proteins, antioxidants, and mineral nutrients; its production increases by herbivory and decreases in the absence of herbivory. Extrafloral nectar has been associated to protective ants, which have the ability to defend their food sources. Increases in extrafloral nectar production augment the numbers of protective ants. In Catalpa bignonioides and Fabaceae family, extrafloral nectar attracts mites, ladybird beetles, wasp, lacewing larvae, and spiders.

Another influent Factors in Plant Defense

The composition and dynamics of the insect community that interacts with plants are influenced by plant traits such as chemistry, physiology, and morphology, which have a genetic basis. Plant traits may affect the sizes of herbivores and therefore the sizes of parasitoids that develop in the herbivores and even the sizes of hyperparasitoids.

Induction of defense timing was examined by Bixenmann and collaborators in Inga genus using lepidopteran larvae on young leaves. While young leaves are expanding, they are tender and high in protein, the two traits that make them a target for herbivores, receiving 70% of the leaf's lifetime herbivore damage despite being vulnerable for only few weeks. Once leaves reach their full size, they rapidly toughen, and rates of herbivore drop to almost zero. The amount of damage, the timing, and the identity of damage agent impact directly induced responses. When increasing leaf area removed in Phaseolus lunatus, extrafloral nectar production, and ant recruitment decreases significantly, then extrafloral nectar production is inversely correlated with leaf area and therefore with the amount of intact photosynthetic surface.

Herbivory risk depends not only on the traits of an individual plant but also on those of neighboring plants. In that sense, the "associational effects" may mediate the local frequency of the density dependence of herbivory.

Volatile organic compounds (VOCs), such as aldehydes, alcohols, esters, and terpenoids, are released from plant flowers, vegetative parts or roots to attract pollinators and predators, repel herbivores, and communicate between or within plants. When a plant is attacked, it is able to communicate with other plants and alert them of a possible future attack; thereby, the alerted plants will respond stronger once attacked. For example, when molasses grass, Melinis minutiflora, was planted in a maize field, the herbivore damage decreased. The grass emits a compound in response to caterpillar damage to attract parasitoids, and the amount of caterpillar in a maize decreased by parasitoids, after induction of JA to release more VOCs.

The perception of herbivory by plants involved not only mechanical injury to plant and the presence of herbivore-derived elicitors released during feeding but also the presence of microbes associated with the herbivore. Microbial symbionts can influence their hosts including providing nutrition, digestion, and detoxifying toxins; insect symbionts have a role in mediating plant defenses. Different microbes in insects may have species-specific effects on different host plants, specifically herbivores' microbiota are perceived by plants during herbivory and thus may alter the outcome plant responses.

Plant Defenses against Herbivores and Fitness

Insects find and select their host plants and deal with plant defenses, as well as herbivores modify plant phenotypes. However, plants interact with multiple attackers and interact at different levels of biological organization.

Herbivory affects the expression of floral traits, plant-pollinator interactions, and costs-benefits to controlling reproductive systems and defense strategies. Plant-herbivore interaction promotes myriad defenses that protect plants from damage. In recent years, it has been considered whether reproductive traits and antiherbivore defenses are interdependent as a result of pollinator- and herbivore-mediated selection. Floral traits are most likely to affect susceptibility to herbivores. There are pollinating herbivores, which when adult insects pollinate the plants their larvae use as host, for example, figs and fig wasps, the larvae feed directly on ovules and developing seeds. A diversity of floral traits influences the susceptibility of plants to herbivores; for example, taller inflorescences often result in greater herbivory, phenology also affects herbivory risk, and plants that flower early or late typically receive less damage than plants that flower during peak flowering.

On the other hand, inbreeding can produce individuals with reduced fitness, but inbred plants are more susceptible to herbivores than outbreds. In horsenettle (Solanum carolinense L.), the tobacco hornworm caterpillars (Manduca sexta L.) preferred to feed on inbred plants, and the females oviposited more frequently on inbred plants compared to outbreds.

Inbreeding in horsenettle causes significant reduction in the plant's induced defense responses and resistance to herbivory. The predilection for inbred plants exhibited by insects suggests that they are gaining fitness benefits by choosing inbred host plants, regulated by insect herbivore growth, oviposition, and flight capacity. Inbred plants, serve as better host for developing insects could be that inbred plants suffer from a limited ability to unregulate genes in defense biochemical pathways. In the system plant-insect horsenettle-tobacco hornworm suggests that biochemical changes in plant inbreeding can influence in the health of animals at a higher trophic level, particularly in insect herbivores which increases survival, growth, and flight metabolism when nurtured on inbred plants.

Tolerance Traits

There is another plant defense strategy: tolerance. In resistance plant synthesizes structural or chemical traits to minimize herbivore damage, while in tolerance traits reduce the negative effects or herbivore damage.

The traits that maintain or promote plant fitness following damage before or after infestation can confer herbivore tolerance, and they are grouped in those that alter physiological process like photosynthesis and growth, phenology, and nutrient storage. In many plant species, partial defoliation leads to increased photosynthetic rate in the remaining plant tissues, but is not universal. Delayed growth, flower, and fruit production following herbivore damage could promote herbivore tolerance by postponing plant development until the threat of attack has passed.

Roots eaten by insect herbivores exhibit extensive regrowth, in density and quantity. The former might be caused by additional lignification that could increase the toughness of the roots.

Mechanisms involved in increased tolerance are i increased net photosynthetic rate after damage, ii high relative growth rates, iii increased branching or tillering after release of apical dominance, iv preexisting high levels of carbon storage in roots for allocation to aboveground reproduction, and v ability to shunt carbon stores from roots to shoots after damage. The evolution of tolerance can promote an apparently mutualistic relationship between plant and herbivore populations.

Example Conifer Plant Defense against Bark Beetle

Now, we examined how different responses can be used by Pinus genera to limit damage causes by attack of bark beetle, one of the principal plagues that affect Pinus populations.

Most herbivores are insects that feed on plants in various forms, for example, they adopt different feeding strategies throughout their life cycle and can feed both external [leaf buds or flowers] and internal structures of the plant [miners, stem borers, gillnet].

Unlike other herbivores such as mammals, insects commonly feed on the leaves and other parts of the mature plant typically do not cause the death of the plant; as for insects to kill the plant, they will require much time. Thus, the relationship between herbivorous insects and plants is more like the host-parasite than predator-prey relationship. Plants for their part have not become passive victims of herbivorous insects as they have been able to produce special metabolites and toxic proteins, which serve as repellents or have antinutritional effects for their attackers. However, herbivorous insects successfully consume plant material, overcoming the complex set of defenses of plant. Moreover, unlike other herbivores, insects are much more specialized, because they can feed exclusively from a plant species or a limited number of them. Therefore, it is necessary to understand the relationship between herbivorous insects and their host plants from biochemical, ecological, behavioral, physiological, and genetic aspects, including the ways in which insects can affect the abundance and distribution of plant species.

Herbivory and Regulation of Plant Populations

Herbivorous insects usually cause reduced growth, fertility, and even the survival of plants; some plants can counter or overcompensate significant amounts of damage in general; however, the insect damage as a group causes a multiple effect and simultaneously in succession with additive effects and multiplicative on the plant fitness, which results in a significant impact on the abundance of plants, distribution, or population dynamics.

The role of herbivorous insects in the regulation of plant populations and dynamics of communities has been poorly documented; most studies have focused mainly on explaining the role of herbivorous insects' native as agents that limit the distribution of its plant host. However, it has been possible to distinguish that the effects of herbivorous insects on plants may differ depending of the different scenarios under which the interaction takes place as in the case of herbivorous insects (bark beetles) and pines.

On the one hand, if the evolutionary success involves adaptive radiation and overtime, the species survive and expand their geographical distribution, and then pines (Pinus sp.) can be considered successful, because they form the largest genus of conifers in the Pinaceae family. The pine group consists of more than 100 species, many subspecies, and varieties. Although mainly distributed in temperate regions of the northern hemisphere, pines also occupy other habitats and climates.

Moreover, the great success of the pines can be attributed to their defense strategies against herbivorous insects or parasites. For its wide distribution and its prolonged generational cycles, ranging from decades to more than 4000 years such as Pinus longaeva, pines are subject to deal with a wide range of attackers at which they have developed along its evolution complex defense mechanisms.

The basic defense strategy of conifers including pines is both morphological structures [physical barriers] and chemical mechanisms. Physical barriers are formed by static structures such as lignified cells, calcium oxalate crystals, or hard foliage; they act primarily against herbivores, ovipositors, and defoliating insects. The bark of the trunk on his part is of particular interest because it forms the first barrier against herbivorous insects such as bark beetles, whose evolution has specialized to kill the tree. Then, conifers produce a plethora of chemical defenses where the most important are phenolic compounds and oleoresins which contain numerous terpenoids. Chemical defense mechanisms may be directed against herbivorous insects to prevent oviposition and food or affect their physiology to reduce survival or fecundity.

Defense and resistance strategies of conifers against bark beetles and fungal pathogens.

Conifers throughout their life cycle face the challenges of a variety of organisms cycle, conifers face the challenges of a variety of organisms, the more severe are the bark beetle and fungal pathogens associated. Conifer defenses against insects and pathogens that infect the trunk are classified as constitutive and induced.

Constitutive Defense Systems

Mechanisms that produce a stable set of structural defenses (cells and resin canals), toxic chemicals such as phenolics and terpenes, and mechanical properties of the cortex (suberized layers of cells and lignified oxalate crystal calcium) are permanent. The constitutive systems are defenses with great resilience against a number of organisms trying to penetrate the cortex during the history of the tree and against common secondary invasions of opportunistic organisms. The constitutive defenses are of two basic types:

1. Mechanical defenses: Structural elements that provide hardness or thickness to tissues and inhibit mastication or piercing in the bark. Impregnating plant tissues with polymers such as suberin and lignins can add resistance to the mechanical properties against penetration, degradation, and ingestion/mastication by insects.

2. Chemical defenses: Formed by chemical compounds stored, like phenolics, terpenoids, and alkaloids, and released under attack. Antinutritive defenses include chemical, toxins, defensive proteins, enzymes, and resin deposits that can flow to repel or physically trap small organisms. These defenses are scattered in the tissues of the bark [periderm, cortex, and secondary phloem]. The constitutive strategies vary depending on the physical or chemical nature of defense and its distribution within the bark and trunk.

Periderm Defenses

Periderm forms a permeable barrier for controlling the gas exchange in the trunk and is the first line of defense against biotic and abiotic factors. It is characterized by the presence of multiple layers of cells, most of which are dead, are also structurally and chemically different, and have lignified or suberized its walls. Cells may contain high amounts of phenolic compounds, and one or more layers have encrusted calcium oxalate crystals. These mechanical defenses (hard walls lignified, crystallization, and suberization) provide a hydrophobic barrier, combined with the chemical properties of the phenolic compounds and form a multifunctional barrier against the external environment. However, the periderm is not a continuous barrier, due to the presence of lenticels to allow gas exchange at the surface, although it is not an open system that may allow entry of invading organisms as in the case of small bark beetles (Pityogenes chalcographus) in Picea abies.

Cortex Defenses

The cortex is formed during the early development of the stem, so it is an important general barrier, especially during the early development of the stem. It remains alive for several years during the secondary growth and contains high amounts of phenolic compounds within vacuoles of cortical parenchyma; in many Pinaceae, the cortex has axial duct resins, which participate in defense, although its function is replaced by the secondary phloem.

Secondary Phloem Defenses

The secondary phloem is the most important site of constitutive defense mechanisms of conifers and is made up of phenolic bodies, sclerenchyma, and calcium oxalate crystals; the relative amount of these components varies considerably between species. A fourth constitutive strategy of defense in certain taxa as Pinaceae is the production of resin structures comprising radial ducts extended from xylem, axial ducts, blisters, and resin cells. The amount and combination of each of these components define defense strategies. In the secondary phloem, there are specialized structures, such as phenolic bodies, sclerenchyma, calcium oxalate crystals, and resins.

The phenolic bodies are parenchymal cells of the axial phloem, also called polyphenolic parenchymal cells [PP cells], specializing in the synthesis and storage of phenolic compounds, making nonedible tissues or antifungal capacity. Different species produce different phenolic compounds depending on the type of organisms that commonly attack, so that the relative resistance to pathogens may be due in part to the type of phenolic compounds they produce.

Moreover, the PP cells are responsible for responses of induced defense and, even when they have thickened walls, allow the exchange of axial and tangential information and signaling for defense because they contain lots of plasmodesmata. The PP cells represent a very dynamic component in defense strategies in conifers and are most abundant in the secondary phloem. Another important feature of the PP cells located along the radial ducts parenchyma is that they are an important site that stores starch and/or lipids, which are considered the target for bark beetles and fungi; however, the presence of phenolic compound constituent allows cells to protect themselves and prevent the penetration of fungi into the area of the cambium. In any case, the layers of PP cells form a sieve maintaining the physical and chemical resistance to prevent penetration into the cortex.

Another important tissue with mechanical function is the sclerenchyma, which is common in the bark of conifers; quantity and type vary among taxa. It consists in cells with thickened lignified secondary wall, which are known as "stone cells" because they are high hardness cells or sclereids, so they can serve as structural element and mechanical defense. This organization is massive and irregular in many Pinaceae or organized form rows as in the case of Taxaceae. Their physical strength can detain predation or perforation of the bark by insects forming a screen of dead cells that progressively collapse under pressure of new layers of inner cells.

The crystals of calcium oxalate formed are stored intracellularly in the secondary phloem of conifers, particularly in Pinaceae, and represents a defense mechanism because the physical nature of the crystals and their relative abundance could imply a role in deterring penetration bark or chewing by herbivores. However, being chemically inert, it is unlikely to have any effect on fungal attack.

One of the common deposits in plants is crystals of calcium oxalate, and its role in most of them is the regulation of calcium; however, also they have secondary functions of defense. In Pinaceae the calcium oxalate crystals embedded within the phenolic bodies in PP cells vacuoles present typically form scattered axial lines crystallized cells. The combination of several layers of fibers and dense encrustation with crystals can provide a powerful defense against bark beetles.

One of the main constitutive defenses is resins, particularly for Pinaceae. The resin production and storage structures for this include radial resin ducts, axial ducts or channels, blisters, and resin cells. Ducts and blisters have a coating epithelial cell enriched by plastids that synthesize terpenoid resins and secreted into the extracellular lumen, which is accumulated under pressure. After injuries are caused by damage from invading organism, the pressed resin is released and may expel the invading organism from the bark and catch it thanks to its sticky consistency or kill it because of its toxic nature. Volatile resin components evaporate and nonvolatile crystallize to sterilize and seal the damaged region effectively. It has been shown that the resin is an effective defense against insect bark borers.

Secondary Xylem Defenses

Secondary xylem is a general system of defense in trunk, which is involved in the synthesis and storage of resin and phenolic compounds and other secondary products such as lignins, and provides a defense against wood-rotting fungi and other organisms. The constituent axial ducts of resin found in the xylem of some conifers can contribute to resin flow when connected to the radial ducts that traverse the xylem and phloem.

Induced Defense Systems

Induced defense system or responses due to herbivore attack involves the synthesis "de novo" or activation of a wide range of chemical defenses, including terpenoids, phenolic compounds, PR proteins, reactive oxygen species, and enzymes. The induced defense system can act against a current infection presenting a hypersensitive response and local resistance or against future infections or attacks by bark beetles generating responses with acquired resistance.

Induced structural defences- Structural defenses in bark are important, because they improve the overall defense capability of the plant; these are diverse and include structural changes and synthesis of chemical and biochemical agents. They are a combination of responses apparently targeting specific organisms, including the general increase in hypersensitivity responses, aimed at limiting the spread of detected damage and isolating the invading organism, repairing damaged tissues, and limiting the attack or later invasion of opportunistic organisms. In addition, long term results in acquired resistance. Among these structural defenses are hypersensitive response, callus tissue formation, and scarring in the periderm.

Hypersensitive Response

Damage produces a hypersensitive response in the plant, which quickly stops invading organisms sacrificing a small piece of tissue. The hypersensitive response occurs locally at the site of infection or attack, producing reactive oxygen species causing rapid cell death, which tries to stop organisms such as pathogenic fungi, bacteria, and virus killing only the damaged plant tissue that has been attacked.

Callus Tissue Formation

A more generalized response in the case of wounds in plants is the formation of callus tissue that can subsequently lignify, suberize, or impregnate phenolic compounds to provide a barrier, part of the wound periderm. This reaction provides protection against new intrusions and blocking an organism such as a fungal pathogen. The callus can also repair damaged tissues so that its functions can be restored.

Scarring in the Periderm

Periderm scars are produced around damaged regions of the cortex, which cause activation of the PP cells of the secondary phloem, which begin to divide to form new tissue. Periderm scar acts as a wall that essentially isolates the damaged area preventing the supply of nutrients to the wound area, which eventually dies if not already dead by the attack of an invading organism. These scars also have permanent effects of tissue repair and generally are formed within the limits of induced injuries by bark beetles or fungal attacks in the trunks of conifers or well around any damaged tissue.

Induced chemical defences - While the constituent chemical defenses are generally nonselective for pest species, induced chemical defenses can be broad-spectrum and specific components. Chemical defenses are extremely diverse and therefore cover a wide range of pests. Nonprotein chemicals, such as products of the phenylpropanoid routes (phenolic) and isoprenoids (terpenoids resin) products, as well as alkaloids can have potent effects on invading organisms.

These compounds are produced more rapidly than protein-based defense because the path usually exists in tissues and only requires activation. However, some of the biochemical pathways are created "de novo" in the tissues.

Another advantage of these chemical defenses is often effective against a wide range of organisms and thus may delay an attack, while recognition mechanisms come into play to identify the organism and then activate specific defenses against herbivore. Among chemical compound induced by herbivores in conifers are phenols, resin terpenoids, and proteins.

Phenolic Compounds

Phenolic compounds are abundant in the bark of conifers, mainly in the PP cells. Both phenolic compounds and tannins act as antifungal agents and block hydrolytic enzymes secreted by invading organisms, thereby inhibiting its progress in tissues. By joining amino acids and proteins disturbed by plant tissues, phenolics and tannins reduce the nutritional value for attackers while coupling to digestive enzymes in the intestine decreases the ability to digest plant tissues. The wounds of the plant or invading organisms in the cortex activate PP cells, including cell expansion and accumulation of a higher amount of phenolic compounds. Generally, the induced phenolic compounds are more toxic or more specific to an invading organism than the constituent phenols, whereby the conversion of polyphenolic compounds to soluble phenolic compounds during an attack adds to the defense capacity; evidence of this is the reduction of polyphenols in vacuoles of intact cells PP near the region of attack.

Resin Terpenoids

Resin terpenoid production is induced by the attack of organisms. During and after attack, the resin flow in the wound can be quite extensive, especially in the Pinaceae. Part of this resin is stored in the structures that produce, while the constituent ducts can be activated to produce resin.

Within the first 2–3 weeks of the attack, the new resin ducts are induced to produce, being considered as traumatic resin ducts, and the resin forming these ducts can be different from the constitutive resin. In Pinaceae and some other groups of conifers, traumatic ducts are formed in the xylem and interconnected with the radial ducts phloem. However, some species of conifers are induced to form more traumatic ducts in the phloem and the xylem. Regardless of their origin, the end result of the development of traumatic resin ducts is to increase the formation and accumulation of resin and increase its flow. The increased flow helps to kill or expel the invaders and to seal the wound and resin-soaked regions of the bark and wood making them more resistant to microbial activity. Furthermore, it has been found that traumatic ducts can confer acquired resistance to subsequent attacks and the resin in traumatic ducts may be more toxic through changes of terpenoids or addition of phenolic compounds.

Proteins

Chemical defenses of the trees based on proteins include enzymes such as chitinases and glucanases that may degrade components of invading organisms and toxic proteins such as porins, lectins, and enzyme inhibitors such as proteinases and amylase. Inhibiting enzymes interfere with the ability of the invading organism to use resources from invaded tissue. Other induced enzymes such as peroxidases and laccases can do more resistant cell walls through crisscrossed reactions or promotion of lignification or well included affecting invader organism. The protein-based defenses can be highly specific for certain organisms. For example, in Norway spruce, there are chitinases as a large family of proteins, but only a small subset of them can be regulated during the attack by a specific pathogenic fungus, and it is presumed that these are effective against the wall cell of this organism. In general, chemical defenses induced mechanical follow a pattern similar to the induced structural defense, such as overlapping of multiple strategies. The production of a toxic cocktail with various chemical components maximizes the potential to stop or destroy an aggressive or virulent invading organism, in contrast to a more conservative production of one or few directed defenses.

Remark Defense System importance in Conifers

Multiple overlaying of structures and defense systems provides an efficient barrier against a wide range of possible attacks of organisms. However, conifers remain susceptible to certain organisms that have evolved strategies to overcome the defenses or avoid them. Nevertheless, the remarkable longevity of various species of conifers is a testament to the success of their defense strategies.

The first line of defense of the plant is given by a mechanical resistance to attack, due to the hardened cells either by thickening the walls or storing different compounds like calcium oxalate crystals that are joined to form a screen of high hardness. This first defense system is effective against most of the organisms that can attack the tree; however, bark beetles usually manage to overcome this barrier, bringing with them pathogenic fungi.

After that penetrate the bark beetles, thanks to its powerful masticatory apparatus, tree active chemical defense mechanisms, in which the phenolic bodies, resin and some proteins may be directed mainly beetles as organisms that are directly attacking the tree; however, these compounds also have an effect on fungi. Another unspecific compound may function to attack bark beetles as in the case of some proteins and calcium oxalate crystals during the attack the hypersensitive response is activated, the formation of callous bodies and interaction with proteins and enzymes which are directed primarily by fungal attack. Also, answers that could be used for both bark beetles and fungi, as in the case of periderm scars, phenolic compounds, and terpenoids, can be triggered. But nevertheless, together, the beetle and the fungus can gradually block the tree's defenses, weakening to lead to death.

Plant Stress

Plant stress is a state where a plant is growing in non-ideal growth conditions and has increased demands put on it. Plant stress refers to any unfavorable condition or substance that affects a plant's metabolism, reproduction, root development, or growth. Plant stress can come in different forms and durations. Some plant stressors are naturally occurring, like drought or wind, while others may be the result of human activity, like over irrigation or root disturbance.

Plant stress can be divided into two primary categories. Abiotic stress is a physical (e.g., light, temperature) or chemical insult that the environment may impose on a plant. Biotic stress is a biological insult, (e.g., insects, disease) to which a plant may be exposed during its lifetime. Some plants may be injured by a stress, which means that they exhibit one or more metabolic dysfunctions. If the stress is moderate and short term, the injury may be temporary and the plant may recover when the stress is removed. If the stress is severe enough, it may prevent flowering, seed formation, and induce senescence that leads to plant death. Such plants are considered to be susceptible. Some plants escape the stress altogether, such as ephemeral, or short-lived, desert plants.

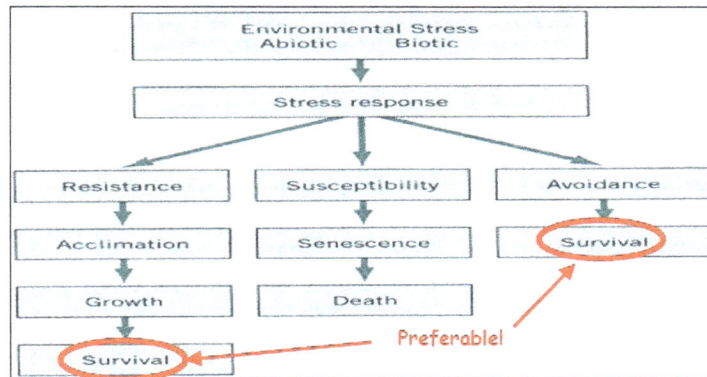

Figure: The effect of environmental stress on plant survival.

Ephemeral plants germinate, grow, and flower very quickly following seasonal rains. They thus complete their life cycle during a period of adequate moisture and form dormant seeds before the onset of the dry season. In a similar manner, many arctic annuals rapidly complete their life cycle during the short arctic summer and survive over winter in the form of seeds. Because ephemeral plants never really experience the stress of drought or low temperature, these plants survive the environmental stress by stress avoidance. Avoidance mechanisms reduce the impact of a stress, even though the stress is present in the environment. Many plants have the capacity to tolerate a particular stress and hence are considered to be stress resistant. Stress resistance requires that the organism exhibit the capacity to adjust or to acclimate to the stress.

Stress resistance requires that the organism exhibit the capacity to adjust or to acclimate to the stress. A plant stress usually reflects some sudden change in environmental condition. However, in stress-tolerant plant species, exposure to a particular stress leads to acclimation to that specific stress in a time-dependent manner. Thus, plant stress and plant acclimation are intimately linked with each other. The stress-induced modulation of homeostasis can be considered as the signal for the plant to initiate processes required for the establishment of a new homeostasis associated with the acclimated state. Plants exhibit stress resistance or stress tolerance because of their genetic capacity to adjust or to acclimate to the stress and establish a new homeostatic state over time. Furthermore, the acclimation process in stress-resistant species is usually reversible upon removal of the external stress.

The establishment of homeostasis associated with the new acclimated state is not the result of a single physiological process but rather the result of many physiological processes that the plant integrates over time, that is, integrates over the acclimation period. Plants usually integrate these physiological processes over a short-term as well as a long-term basis. The *short-term processes* involved in acclimation can be initiated within seconds or minutes upon exposure to a stress but may be transient in nature. That means that although these processes can be detected very soon after the onset of a stress, their activities also disappear rather rapidly. As a consequence, the lifetime of these processes is rather short. In contrast, *long-term processes* are less transient and thus usually exhibit a longer lifetime. However, the lifetimes of these processes overlap in time such that the short-term processes usually constitute the initial responses to a stress while the long-term processes are usually detected later in the acclimation process. Such a hierarchy of short- and long-term responses indicates that the attainment of the acclimated state can be considered a complex, time-nested response to a stress. Acclimation usually involves the differential expression of specific sets of genes associated with exposure to a particular stress. The remarkable capacity to *regulate gene expression* in response to environmental change in a time-nested manner is the basis of plant plasticity.

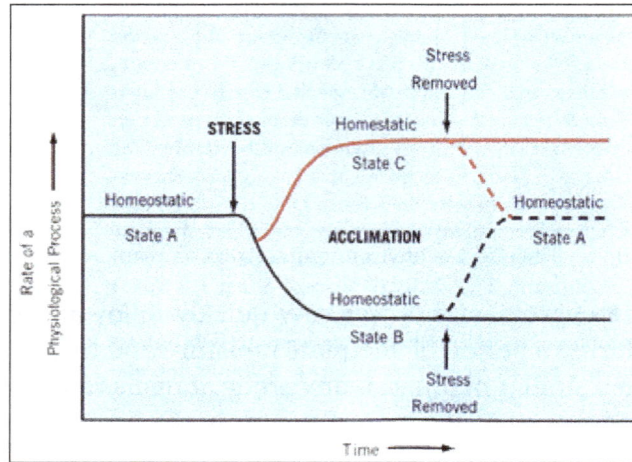

A schematic relationship between stress and acclimation.

Adaptation and Phenotypic Plasticity

Plants have various mechanisms that allow them to survive and often prosper in the complex environments in which they live. Adaptation to the environment is characterized by genetic changes in the entire population that have been fixed by natural selection over many generations. In contrast, individual plants can also respond to changes in the environment, by directly altering their physiology or morphology to allow them to better survive the new environment. These responses require no new genetic modifications, and if the response of an individual improves with repeated exposure to the new environmental condition then the response is one of acclimation. Such responses are often referred to as phenotypic plasticity, and represent nonpermanent changes in the physiology or morphology of the individual that can be reversed if the prevailing environmental conditions change.

Individual plants may also show phenotypic plasticity that allows them to respond to environmental fluctuations. In addition to genetic changes in entire populations, individual plants may also show phenotypic plasticity; they may respond to fluctuations in the environment by directly altering their morphology and physiology. The changes associated with phenotypic plasticity require no new genetic modifications, and many are reversible. Both genetic adaptation and phenotypic plasticity can contribute to the plant's overall tolerance of extremes in their abiotic environment. As a consequence, a plant's physiology and morphology are not static but are very dynamic and responsive to their environment. The ability of biennial plants and winter cultivars of cereal grains to survive over winter is an example of acclimation to low temperature. The process of acclimation to a stress is known as hardening and plants that have the capacity to acclimate are commonly referred to as hardy species. In contrast, those plants that exhibit a minimal capacity to acclimate to a specific stress are referred to as nonhardy species.

Imbalances of Abiotic Factors have Primary and Secondary Effects on Plants

Plants may experience physiological stress when an abiotic factor is deficient or in excess (referred to as an imbalance). The deficiency or excess may be chronic or intermittent. Abiotic conditions to which native plants are adapted may cause physiological stress to non-native plants. Most agricultural crops, for example, are cultivated in regions to which they are not highly adapted. Field crops are estimated to produce only 22% of their genetic potential for yield because of suboptimal climatic and soil conditions.

Imbalances of abiotic factors in the environment cause *primary and secondary effects* in plants. Primary effects such as reduced water potential and cellular dehydration directly alter the physical and biochemical properties of cells, which then lead to secondary effects. These secondary effects, such as reduced metabolic activity, ion cytotoxicity, and the production of reactive oxygen species, initiate and accelerate the disruption of cellular integrity, and may lead ultimately to cell death. Different abiotic factors may cause similar primary physiological effects because they affect the same cellular processes. This is the case for water deficit, salinity, and freezing, all of which cause reduction in hydrostatic pressure (turgor pressure, Ψp) and cellular dehydration. Secondary physiological effects caused by different abiotic imbalances may overlap substantially. It is evident that imbalances in many abiotic factors reduce cell proliferation, photosynthesis, membrane integrity, and protein stability, and induce production of *reactive oxygen species (ROS)*, oxidative damage, and cell death.

The Light-dependent Inhibition of Photosynthesis

As photoautotrophs, plants are dependent upon – and exquisitely adapted to – visible light for the maintenance of a positive carbon balance through photosynthesis. Higher energy wavelengths of electromagnetic radiation, especially in the ultraviolet range, can inhibit cellular processes by damaging membranes, proteins, and nucleic acids. However, even in the visible range, irradiances far above the light saturation point of photosynthesis cause high light stress, which can disrupt chloroplast structure and reduce photosynthetic rates, a process known as photoinhibition.

Photoinhibition by high light leads to the production of destructive forms of oxygen.

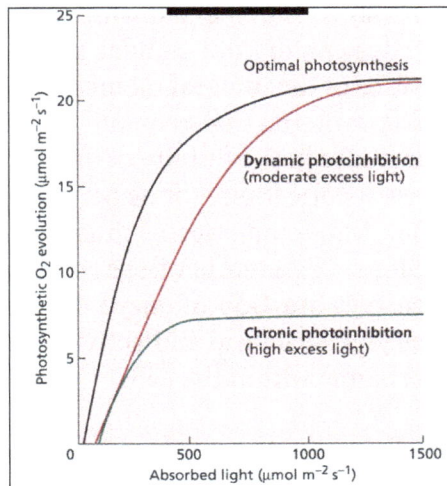

Changes in the light-response curves of photosynthesis caused by photoinhibition.

Excess light excitation arriving at the PSII reaction center can lead to its inactivation by the direct damage of the D1 protein. Excess absorption of light energy by photosynthetic pigments also produces excess electrons outpacing the availability of $NADP^+$ to act as an electron sink at PSI. The excess electrons produced by PSI lead to the production of reactive oxygen species (ROS), notably superoxide (O_2^-). Superoxide and other ROS are low-molecular-weight molecules that function in signaling and, in excess, cause oxidative damage to proteins, lipids, RNA, and DNA. The oxidative stress generated by excessive ROS destroys cellular and metabolic functions and leads to cell death.

Temperature Stress

Mesophytic plants (terrestrial plants adapted to temperate environments that are neither excessively wet nor dry) have a relatively narrow temperature range of about 10 °C for optimal growth and development. Outside of this range, varying amounts of damage occur, depending on the magnitude and duration of the temperature fluctuation. Temperature Stress can be divided into three categories high temperatures, low temperatures above freezing, and temperatures below freezing. Most actively growing tissues of higher plants are tillable to survive extended exposure to temperatures above 45 °C or even short exposure to temperatures of 55 °C or above. However, nongrowing cells or dehydrated tissues (e.g., seeds and pollen) remain viable at much higher temperatures. Pollen grains of some species can survive 70 °C and some dry seeds can tolerate temperatures as high as 120 °C.

Most plants with access to abundant water are able to maintain leaf temperatures below 45 °C by evaporative cooling, even at elevated ambient temperatures. However, high leaf temperatures combined with minimal evaporative cooling causes heat stress. Leaf temperatures can rise to 4 to 5 °C above ambient air temperature in bright sunlight near midday, when soil water deficit causes partial stomatal closure or when high relative humidity reduces the gradient driving evaporative cooling. Increases in leaf temperature during the day can be more pronounced in plants experiencing drought and high irradiance from direct sunlight.

Temperature Stress can Result in Damaged Membranes and Enzymes

Plant membranes consist of a lipid bilayer interspersed with proteins and sterols, and any abiotic factor that alters membrane properties can disrupt cellular processes. The physical properties of the lipids greatly influence the activities of the integral membrane proteins, including H^+ pumping Pases, carriers, and channel-forming proteins that regulate the transport of ions and other solutes. High temperatures cause an increase in the fluidity of membrane lipids and a decrease in the strength of hydrogen bonds and electrostatic interactions between polar groups of proteins within the aqueous phase of the membrane. High temperatures thus modify membrane composition and structure, and can cause leakage of ions. High temperatures can also lead to a loss of the three-dimensional structure required for correct function of enzymes or structural cellular components, thereby leading to loss of proper enzyme structure and activity. Misfiled proteins often aggregate and precipitate, creating serious problems within the cell.

Temperature Stress can Inhibit Photosynthesis

Photosynthesis and respiration are both inhibited by temperature stress. Typically, photosynthetic rates are inhibited by high temperatures to a greater extent than respiratory rates. Although chloroplast enzymes such as rubisco, rubisco activase, NADP-G3P dehydrogenase, and PEP carboxylase become unstable at high temperatures, the temperatures at which these enzymes began to denature and lose activity are distinctly higher than the temperatures at which photosynthetic rates begin to decline. This would indicate that the early stages of heat injury to photosynthesis are more directly related to changes in membrane properties and to uncoupling of the energy transfer mechanisms in chloroplasts.

This imbalance between photosynthesis and respiration is one of the main reasons for the deleterious effects of high temperatures. On an individual plant, leaves growing in the shade have a lower temperature compensation point than leaves that are exposed to the sun (and heat). Reduced photosynthate production may also result from stress-induced stomatal closure, reduction in leaf canopy area, and regulation of assimilate partitioning.

Freezing Temperatures cause Ice Crystal Formation and Dehydration

Freezing temperatures result in intra- and extracellular ice crystal formation. Intracellular ice formation physically shears membranes and organelles. Extracellular ice crystals, which usually form before the cell contents freeze, may not cause immediate physical damage to cells, but they do cause cellular dehydration. This is because ice formation substantially lowers the water potential (Ψw) in the apoplast, resulting in a gradient from high Ψw in the symplast to low Ψw in the apoplast. Consequently, water moves from the symplast to the apoplast, resulting in cellular dehydration. Cells that are already dehydrated, such as those in seeds and pollen, are relatively less affected by ice crystal formation. Ice usually forms first within the intercellular spaces and in the xylem vessels, along which the ice can quickly propagate. This ice formation is not lethal to hardy plants, and the tissue recovers fully if warmed. However, when plants are exposed to freezing temperatures for an extended period, the growth of extracellular ice crystals leads to physical destruction of membranes and excessive dehydration.

Imbalances in Soil Minerals

Imbalances in the mineral content of soils can affect plant fitness either indirectly, by affecting plant nutritional status or water uptake, or directly, through toxic effects on plant cells.

1. Soil mineral content can result in plant stress in various ways:

Several anomalies associated with the elemental composition of soils can result in plant stress, including high concentrations of salts (e.g., $Na+$ and $Cl-$) and toxic ions (e.g., As and Cd), and low concentrations of essential mineral nutrients, such as Ca^{2+}, Mg^{2+}, N, and P. The term salinity is used to describe excessive accumulation of salt in the soil solution. Salinity stress has two components: nonspecific osmotic stress that causes water deficits, and specific ion effects resulting from the accumulation of toxic ions, which disturb nutrient acquisition and result in cytotoxicity. Salt-tolerant plants genetically adapted to salinity are termed *halophytes*, while less salt-tolerant plants that are not adapted to salinity are termed glycophytes.

2. Soil salinity occurs naturally and as the result of improper water management practices:

In natural environments, there are many causes of salinity. Terrestrial plants encounter high salinity close to the seashore and in estuaries where seawater and freshwater mix or replace each other with the tides. The movement of seawater upstream into rivers can be substantial, depending on the strength of the tidal surge. Far inland, natural seepage from geologic marine deposits can wash salt into adjoining areas. Evaporation and transpiration remove pure water (as vapor) from the soil, concentrating the salts in the soil solution. Soil salinity is also increased when water droplets from the ocean disperse over land and evaporate.

Human activities also contribute to soil salinization. Improper water management practices associated with intensive agriculture can cause substantial salinization of croplands. In many areas of the world, salinity threatens the production of staple foods. Irrigation water in semiarid and arid regions is often saline. Only halophytes, the most salt-tolerant plants, can tolerate high levels of salts. Glycophytic crops cannot be grown with saline irrigation water.

Saline soils are often associated with high concentrations of NaCl, but in some areas Ca^{2+}, Mg^{2+}, and SO_4^- are also present in high concentrations in saline soils. High Na^+ concentrations that occur in sodic soils (soils in which Na^+ occupies 10% of the cation exchange capacity) not only injure plants but also degrade the soil structure, decreasing porosity and water permeability. Salt incursion into the soil solution causes water deficits in leaves and inhibits plant growth and metabolism.

3. High cytosolic Na^+ and Cl^- denature proteins and destabilize membranes:

The most widespread example of a specific ion effect is the cytotoxic accumulation of Na^+ and Cl^- ions under saline conditions. Under non-saline conditions, the cytosol of higher plant cells contains about 100 m M K^+ and less than 10 m M Na^+, an ionic environment in which enzymes are optimally functional. In saline environments, cytosolic Na^+ and Cl^- increase to more than 100 mM, and these ions become cytotoxic. High concentrations of salt cause protein denaturation and membrane destabilization by reducing the hydration of these macromolecules. However, Na^+ is a more potent denaturant than K^+.

At high concentrations, apoplastic Na^+ also competes for sites on transport proteins that are necessary for high-affinity uptake of K^+, an essential macronutrient. Further, Na^+ displaces Ca^{2+} from sites on the cell wall, reducing Ca^{2+} activity in the apoplast and resulting in greater Na^+ influx, presumably through nonselective cation channels. Reduced apoplastic Ca^{2+} concentrations caused by excess Na^+ may also restrict the availability of Ca^{2+} in the cytosol. Since cytosolic Ca^{2+} is necessary to activate Na^+ detoxification via efflux across the plasma membrane, elevated external Na^+ has the ability to block its own detoxification.

Developmental and Physiological Mechanisms against Environmental Stress

1. Plants can modify their life cycles to avoid abiotic stress:

One way plants can adapt to extreme environmental conditions is through modification of their life cycles. For example, annual desert plants have short life cycles: they complete them during the periods when water is available, and are dormant (as seeds) during dry periods. Deciduous trees of the temperate zone shed their leaves before the winter so that sensitive leaf tissue is not damaged by cold temperatures. During less predictable stressful events (e.g., a summer of significant but erratic rainfall) the growth habits of some species may confer a degree of tolerance to these conditions. For example, plants that can grow and flower over an extended period (*indeterminate growth*) are often more tolerant to erratic environmental extremes than plants that develop preset numbers of leaves and flower over only very short periods (*determinate growth*).

2. Phenotypic changes in leaf structure and behavior are important stress responses:

Because of their roles in photosynthesis, leaves (or their equivalent) are crucial to the survival of a plant. To function, leaves must be exposed to sunlight and air, but this also makes them particularly vulnerable to environmental extremes. Plants have thus evolved various mechanisms that enable them to avoid or mitigate the effects of abiotic extremes to leaves. Such mechanisms include changes in leaf area, leaf orientation, trichomes, and the cuticle.

Turgor reduction is the earliest significant biophysical effect of water deficit. As a result, turgor-dependent processes such as *leaf expansion* and root elongation are the most sensitive to water deficits. When water deficit develops slowly enough to allow changes in developmental processes, it has several effects on growth, one of which is a limitation of leaf expansion. Because leaf expansion depends mostly on cell expansion, the principles that underlie the two processes are similar. Inhibition of cell expansion results in a slowing of leaf expansion early in the development of water deficits. The resulting smaller leaf area transpires less water, effectively conserving a limited water supply in the soil over a longer period. Altering *leaf shape* is another way that plants can reduce leaf area. Under conditions of water, heat, or salinity extremes, leaves may be narrower or may develop deeper lobes during development. The result is a reduced leaf surface area and therefore, reduced water loss and heat load (defined as amount of heat loss [cooling] required to maintain a leaf temperature close to air temperature). For protection against overheating during water deficit, the leaves of some plants may orient themselves away from the sun. *Leaf orientation* may also change in response to low oxygen availability.

Altered leaf shape can occur in response to environmental changes: leaf from outside (left) and inside (right) of a tree canopy.

3. Plants can regulate stomatal aperture in response to dehydration stress:

The ability to control stomatal aperture allows plants to respond quickly to a changing environment, for example to avoid excessive water loss or limit uptake of liquid or gaseous pollutants through stomata. Stomatal opening and closing is modulated by uptake and loss of water in guard cells, which changes their turgor pressure. Although guard cells can lose turgor as a result of a direct loss of water by evaporation to the atmosphere, stomatal closure in response to dehydration is almost always an active, energy-dependent process rather than a passive one. Abscisic acid (ABA) mediates the solute loss from guard cells that is triggered by a decrease in the water content of the leaf. Plants constantly modulate the concentration and cellular localization of ABA, and this allows them to respond quickly to environmental changes, such as fluctuations in water availability.

4. Plants adjust osmotically to drying soil by accumulating solutes:

Osmotic adjustment is the capacity of plant cells to accumulate solutes and use them to lower Ψw during periods of osmotic stress. The adjustment involves a net increase in solute content per cell that is independent of the volume changes that result from loss of water. The decrease in ΨS (= osmotic potential) is typically limited to about 0.2 to 0.8 MPa, except in plants adapted to extremely dry conditions.

There are two main ways by which osmotic adjustment can take place. A plant may *take up ions* from the soil, or *transport ions* from other plant organs to the root, so that the solute concentration of the root cells increases. For example, increased uptake and accumulation of K+ will lead to decreases in ΨS due to the effect of the potassium ions on the osmotic pressure within the cell. This is a common event in saline areas, where ions such as potassium and calcium are readily available to the plant. The accumulation of ions during osmotic adjustment is predominantly restricted to the vacuoles, where the ions are kept out of contact with cytosolic enzymes or organelles.

When ions are compartmentalized in the vacuole, other solutes must accumulate in the cytoplasm to maintain water potential equilibrium within the cell. These solutes are called *compatible solutes* (or *compatible osmolytes*). Compatible solutes are organic compounds that are osmotically active in the cell, but do not destabilize the membrane or interfere with enzyme function, as high concentrations of ions can. Plant cells can hold large concentrations of these compounds without detrimental effects on metabolism. Common compatible solutes include amino acids such as proline, sugar alcohols such as mannitol, and quaternary ammonium compounds such as glycine betaine.

5. Phytochelatins chelate certain ions, reducing their reactivity and toxicity:

Chelation is the binding of an ion with at least two ligating atoms within a chelating molecule. Chelating molecules can have different atoms available for ligation, such as sulfur (S), nitrogen (N), or oxygen (O), and these different atoms have different affinities for the ions they chelate. By wrapping itself around the ion it binds to form a complex, the chelating molecule renders the ion less chemically active, thereby reducing its potential toxicity. The complex is then usually translocated to other parts of the plant, or stored away from the cytoplasm (typically in the vacuole). Phytochelatins are low-molecular-weight thiols consisting of the amino acids glutamate, cysteine, and glycine, with the general form of (γ-Glu-Cys)nGly. The thiol groups act as ligands for ions of trace elements such as Cd and As. Once formed, the phytochelatin-metal complex is transported into the vacuole for storage.

6. Many plants have the capacity to acclimate to cold temperature:

The ability to tolerate freezing temperatures under natural conditions varies greatly among tissues. Seeds and other partially dehydrated tissues, as well as fungal spores, can be kept indefinitely at temperatures near absolute zero (0 K, or -273 °C), indicating that these very low temperatures are not intrinsically harmful. Hydrated, vegetative cells can also retain viability at freezing temperatures, provided that ice crystal formation can be restricted to the intercellular spaces and cellular dehydration is not too extreme.

Temperate plants have the capacity for *cold acclimation* – a process whereby exposure to low but nonlethal temperatures (typically above freezing) increases the capacity for low temperature survival. Cold acclimation in nature is induced in the early autumn by exposure to short days and nonfreezing, chilling temperatures, which combine to stop growth. A diffusible factor that promotes acclimation, most likely ABA, moves from leaves via the phloem to overwintering stems. ABA accumulates during cold acclimation and is necessary for this process.

7. Plants survive freezing temperatures by limiting ice formation:

During rapid freezing, the protoplast, including the vacuole, may supercool; that is, the cellular water remains liquid because of its solute content, even at temperatures several degrees below its theoretical freezing point. Supercooling is common to many species of the hardwood forests. Cells can supercool to only about -40 °C, the temperature at which ice forms spontaneously. Spontaneous ice formation sets the low-temperature limit at which many alpine and subarctic species that undergo deep supercooling can survive. It may also explain why the altitude of the timberline in mountain ranges is at or near the -40 °C minimum isotherm. Several specialized plant proteins, termed antifreeze proteins, limit the growth of ice crystals through a mechanism independent of lowering of the freezing point of water. Synthesis of these antifreeze proteins is induced by cold temperatures. The proteins bind to the surfaces of ice crystals to prevent or slow further crystal growth.

8. Cold-resistant plants tend to have membranes with more unsaturated fatty acids:

As temperatures drop, membranes may go through a phase transition from a flexible liquid-crystalline structure to a solid gel structure. The phase transition temperature varies with species (tropical species: 10-12 °C; apples: 3-10 °C) and the actual lipid composition of the membranes. Chilling-resistant plants tend to have membranes with more unsaturated fatty acids. Chilling-sensitive plants, on the other hand, have a high percentage of saturated fatty acid chains, and membranes with this composition tend to solidify into a semicrystalline state at a temperature well above 0 °C. Prolonged exposure to extreme temperatures may result in an altered composition of membrane lipids, a form of acclimation. Certain transmembrane enzymes can alter lipid saturation, by introducing one or more double bonds into fatty acids. This modification lowers the temperature at which the membrane lipids begin a gradual phase change from fluid to semicrystalline form and allows membranes to remain fluid at lower temperatures, thus protecting the plant against damage from chilling.

9. A large variety of heat shock proteins can be induced by different environmental conditions:

Under environmental extremes, protein structure is sensitive to disruption. Plants have several mechanisms to limit or avoid such problems, including osmotic adjustment for maintenance of hydration and chaperone proteins that physically interact with other proteins to facilitate protein folding, reduce misfolding and aggregation, and stabilize protein tertiary structure. In response to sudden 5 to 10 °C increases in temperature, plants produce a unique set of chaperone proteins referred to as heat shock proteins (HSPs). Cells that have been induced to synthesize HSPs show improved thermal tolerance and can tolerate subsequent exposure to temperatures that otherwise would be lethal. Heat shock proteins are also induced by widely different environmental conditions, including water deficit, ABA treatment, wounding, low temperature, and salinity. Thus, cells that have previously experienced one condition may gain cross-protection against another.

During mild or short-term water shortage, photosynthesis is strongly inhibited, but phloem translocation is unaffected until the shortage becomes severe. Changes in the environment may stimulate shifts in metabolic pathways. When the supply of O_2 is insufficient for aerobic respiration, roots first begin to ferment pyruvate to lactate through the action of lactate dehydrogenase; this recycles NADH to NAD^+, allowing the maintenance of ATP production through glycolysis. Production of lactate (lactic acid) lowers the intracellular pH, inhibiting lactate dehydrogenase and activating pyruvate decarboxylase. These changes in enzyme activity quickly lead to a switch from lactate to ethanol production. The net yield of ATP in fermentation is only 2 moles of ATP per mole of hexose sugar catabolized (compared with 36 moles of ATP per mole of hexose respired in aerobic respiration). Thus, injury to root metabolism by O_2 deficiency originates in part from a lack of ATP to drive essential metabolic processes such as root absorption of essential nutrients.

Water shortage decreases both photosynthesis and the consumption of assimilates in the expanding leaves. As a consequence, water shortage indirectly decreases the amount of photosynthate exported from leaves. Because phloem transport depends on pressure gradients, decreased water potential in the phloem during water deficit may inhibit the movement of assimilates. The ability to continue translocating assimilates is a key factor in almost all aspects of plant resistance to drought.

References

- Plant-ecology, plant-ecology-traditional-approaches-to-recent-trends: intechopen.com, Retrieved 11 April, 2019

- Soil-plant-interactions, envirobiology: openoregon.pressbooks.pub, Retrieved 21 August, 2019

- Interactions-plant-plant, news-wires-white-papers-and-books, science: encyclopedia.com, Retrieved 20 March, 2019

- Animal-plant-interactions: lifeofplant.blogspot.com, Retrieved 7 May, 2019

- Chemical-plant-defense-against-herbivores, herbivores: intechopen.com, Retrieved 10 June, 2019

- Plant-stress-what-causes-plant-stress-and-how-to-reduce-it: coolplanet.com, Retrieved 31 January, 2019

Chapter 6

Animal Ecology

The branch of ecology which deals with the interactions between animal populations and their environment is known as animal ecology. Some of the different topics studied within this domain are camouflage, parasitism and mutualisms in animals, as well as predator-prey relationships. This chapter discusses in detail these theories and concepts related to animal ecology.

Animal ecology concerns the relationships of individuals to their environments, including physical factors and other organisms, and the consequences of these relationships for evolution, population growth and regulation, interactions between species, the composition of biological communities, and energy flow and nutrient cycling through the ecosystem. From the standpoint of population, the individual organism is the fundamental unit of ecology. Factors influencing the survival and reproductive success of individuals form the basis for under-standing population processes.

Two general principles guide the study of animal ecology. One is the balance of nature, which states that ecological systems are regulated in approximately steady states. When a population becomes large, ecological pressures on population size, including food shortage, predation, disease, tend to reduce the number of individuals. The second principle is that populations exist in dynamic relationship to their environments and that these relationships may cause ecological systems to vary dramatically over time and space. One of the challenges of animal ecology has been to reconcile these different viewpoints.

Populations depend on resources, including space, food, and opportunities to escape from predators. The amount of a resource potentially available to a population is generally thought of as being a property of the environment. As individuals consume resources they reduce the availability of these resources to others in the population. Thus, individuals are said to compete for resources. Larger populations result in a smaller share of resources per individual, which may lead to reduced survival and fecundity. Dense populations also attract predators and provide conditions for rapid transmission of contagious diseases, which generate pressure to reduce population size.

Changes in population size reflect both extrinsic variation in the environment that affects birth and death rates and intrinsic dynamics that result in oscillations or irregular fluctuations in population size. In some situations, the stable state may be a regular oscillation known as a limit cycle. Ecological systems also may switch between alternative stable states, as in the case of populations that are regulated at a high level by food limitation or at a low level by predators or other enemies. Switching between alternative stable states may be driven by changes in the environment.

Population Increase

In the absence of the effects of crowding, all populations have an immense capacity to increase. This capacity may be expressed as an exponential growth rate, which describes the growth of a population in terms of its relative, or percentage, rate of increase, like continuously compounded interest on a bank account. The constant r is often referred to as the Malthusian parameter. For a population growing at an exponential rate, the number of individuals (N) in a population at time t is $N(t) = (0)e^{rt}$ where $N(0)$ is the number of individuals at time 0. Accordingly, the increase in a single time unit is e^r, which is the constant factor by which the population increases during each time period. The rate of increase in the number of individuals is then given by $dN/dt = rN$. The doubling time in years of a population growing exponentially is $t_2 = (ln\ 2)/r$, or roughly $0.69/r$.

Estimated exponential annual growth rates of unrestrained populations range from low values of 0.077 for sheep in Tasmania and 0.091 for Northern elephant seals, to perhaps 1.0 for a pheasant population, 24 for the field vole, 10^{10} for flour beetles in laboratory cultures and 10^{30} for the water flea *daphnia*. Human populations are at the lower end of this range, but a realistic exponential growth rate of 0.03 (or slightly above 3% per year) for some human populations is equivalent to a doubling time of about 23 years and a roughly thousand-fold increase in 230 years. Clearly, no population can maintain such a growth rate for long. (Expansion at the estimated annualized rates just cited for the field vole, flour beetle, and water flea is necessarily utterly fleeting).

The exponential growth rate of a population can be calculated from the schedule of fecundity at age x (b_x) and survival to age l) in a population. These "life table" variables are related to population growth rate by the Euler, or characteristic, equation, whose solution requires matrix methods. When the life table is unchanging for a long period, a population assumes a stable age distribution, which is also an intrinsic property of the life table, and a constant exponential rate of growth. Thus, assuming constant birth and death rates, the growth trajectory of a population may be projected into the future. However, because populations are finite and births and deaths are random events, the expected size of a population in the future has a statistical distribution that may include a finite probability of 0 individuals, that is, extinction. As a general rule, the probability of extinction decreases with increasing population size and increasing excess of births over deaths.

Population Regulation

Balancing the growth potential of all populations are various extrinsic environmental factors that act to slow population growth as the number of individuals increases. High population density depresses the resources of the environment, attracts predators, and, in some cases, results in stress-related reproductive failure or premature death. As population size increases, typically death rates of individuals increase, birth rates decrease, or both. The result is a slower growth rate and a changed, usually older, population. The predominant model used by animal ecologists to describe the relationship of population growth rate to population size (or density) is the logistic equation, in which the exponential growth rate of the population decreases linearly with increasing population size:

Where r_0 is the exponential growth rate of a population unrestrained by density (i.e., whose size is close to 0) and K represents the number of individuals that the environment can support at an equilibrium level, also referred to as the carrying capacity of the environment. Accordingly, the rate of growth of the population is expressed as

Notice that when N < K, the growth rate is positive and the population grows. When N >K, the density-dependent term $(1 - N/K)$ is negative and the population declines. When $N = K$, the growth rate is 0 and a stable, steady-state population size is achieved. This depressing impact of density on the population growth rate is known as negative feedback.

The differential form of the logistic equation may be integrated to provide a function for the trajectory of population size over time.

The curve is sigmoid (S-shaped), with the rate of growth, dN/dt reaching a maximum (the inflection point) at $N = K/2$. Because this is the density at which individuals are added to the population most rapidly, the inflection point also represents the size of the population from which human consumers can remove individuals at the highest rate without causing the population to decline. Thus, the inflection point is also known as the point of maximum sustainable yield.

Density dependence can take on a variety of forms. One of these is a saturation model where the exponential growth rate remains constant and positive until a population completely utilizes a nonrenewable resource such as space, and population growth stops abruptly. The approach of a population to an equilibrium level determined by density-dependent processes can be altered by environmentally induced changes in the intrinsic rate of population growth or in the carrying capacity of the environment.

Difficulties in finding mates and maintaining other social interactions at low densities, including group defense against predators, may also cause the population growth rate to decrease as density declines (Allee effect), and, below a certain density threshold, may even result in population decline to extinction. This type of response is a positive feed-back, one that promotes population instability. For example, after commercial hunting had reduced populations of the passenger pigeon to low levels, the decline in social interactions in this communally nesting species is thought to have doomed it to extinction.

Populations have inherent oscillatory properties that can be triggered by time lags in the response to changing density and which cause populations to fluctuate in a perpetual limit cycle, with alternating population highs and lows. In these cases higher values of r can send a population into unpredictable chaotic behavior, increasing the risk of extinction. In a population with continuous reproduction, regular population cycles occur when there is a lag, often equal to the period of development, in the response of a population to its own density effects on the environment. When the time lag is of period τ, limit cycles develop when $r\tau$ exceeds $\pi/2$, and the period of the cycle is 4 to 5 times τ.

Metapopulations

Most natural populations consist of many subpopulations occupying patches of suitable habitat surrounded by unsuitable environments. Oceanic islands and freshwater ponds are obvious examples. But fragmentation of forest and other natural habitats resulting from clearing land for agriculture or urban development is increasingly creating fragmented populations in many other kinds of habitats. These subpopulations are connected by movement of individuals, and the set of subpopulations is referred to as a metapopulation. Metapopulations have their own dynamics determined by the probabilities of colonization and extinction of individual patches. A set of simple metapopulation models describes changes in the proportion of patches occupied (). When the extinction probability (e) of an individual patch is independent of p, the rate of loss of subpopulations is simply pe. The rate of colonization is proportional to the number of patches that can provide potential colonists and the proportion of empty patches that are available to receive them. Hence, colonization is equal to cp(1-p), where c is the rate of colonization.

The metapopulation achieves a steady state of number of patches occupied when colonization balances extinction, that is pe = cp(1-p), or $\hat{p}$ = 1-e/c. In this model, as long as the rate of colonization exceeds that of extinction, the metapopulation will persist. In more complex models, particularly when the probability of population extinction is reduced by continuing migration of individuals between patches (which keeps the sizes of subpopulations from dropping perilously low), the extinction rate and colonization rate both depend on patch occupancy. In this case, the solution to the metapopulation model has a critical ratio of colonization to extinction, below which patch occupancy declines until the metapopulation disappears. Thus, changes in patch size or migration between patches can cause an abrupt shift in the probability of metapopulation persistence.

Camouflage in Animals

Camouflage, also called cryptic coloration, is a defense or tactic that organisms use to disguise their appearance, usually to blend in with their surroundings. Organisms use camouflage to mask their location, identity, and movement. This allows prey to avoid predators, and for predators to sneak up on prey.

A species' camouflage depends on several factors. The physical characteristics of the organism are important. Animals with fur rely on different camouflage tactics than those with feathers or scales, for instance. Feathers and scales can be shed and changed fairly regularly and quickly. Fur, on the other hand, can take weeks or even months to grow in. Animals with fur are more often camouflaged by season. The arctic fox, for example, has a white coat in the winter, while its summer coat is brown.

The behavior of a species is also important. Animals that live in groups differ from those that are solitary. The stripes on a zebra, for instance, make it stand out. However, zebras are social animals, meaning they live and migrate in large groups called herds. When clustered together, it is nearly impossible to tell one zebra from another, making it difficult for predators such as lions to stalk an individual animal.

A species' camouflage is also influenced by the behavior or characteristics of its predators. If the predator is color-blind, for example, the prey species will not need to match the color of its surroundings. Lions, the main predator of zebras, are color-blind. Zebras' black-and-white camouflage does not need to blend in to their habitat, the golden savanna of central Africa.

Camouflage Tactics

Environmental and behavioral factors cause species to employ a wide variety of camouflage tactics.

Some of these tactics, such as background matching and disruptive coloration, are forms of mimicry. Mimicry is when one organism looks or acts like an object or another organism.

Background matching is perhaps the most common camouflage tactic. In background matching, a species conceals itself by resembling its surroundings in coloration, form, or movement. In its simplest form, animals such as deer and squirrels resemble the "earth tones" of their surroundings. Fish such as flounder almost exactly match their speckled seafloor habitats.

More complex forms of background matching include the camouflage of the walking stick and walking leaf. These two insects, both native to southeast Asia, look and act like their namesakes. Patterns on the edge of the walking leaf's body resemble bite marks left by caterpillars in leaves. The insect even sways from side to side as it walks, to better mimic the swaying of a leaf in the breeze.

Another camouflage tactic is disruptive coloration. In disruptive coloration, the identity and location of a species may be disguised through a coloration pattern. This form of visual disruption causes predators to misidentify what they are looking at. Many butterflies have large, circular patterns on the upper part of their wings. These patterns, called eyespots, resemble the eyes of animals much larger than the butterfly, such as owls. Eyespots may confuse predators such as birds and misdirect them from the soft, vulnerable part of the butterfly's body.

Other species use coloration tactics that highlight rather than hide their identity. This type of camouflage is called warning coloration or aposematism. Warning coloration makes predators aware of the organism's toxic or dangerous characteristics. Species that demonstrate warning coloration include the larva and adult stages of the monarch butterfly. The monarch caterpillar is brightly striped with yellow, black, and white. The monarch butterfly is patterned with orange, black, and white. Monarchs eat milkweed, which is a poison to many birds. Monarchs retain the poison in their bodies. The milkweed toxin is not deadly, but the bird will vomit. The bright coloring warns predator birds that an upset stomach is probably not worth a monarch meal.

Another animal that uses aposematism is the deadly coral snake, whose brightly colored rings alert other species to its toxic venom. The coral snake's warning coloration is so well known in the animal kingdom that other, non-threatening species mimic it in order to camouflage their true identities. The harmless scarlet king snake has the same black, yellow, and red striped pattern as the coral snake. The scarlet king snake is camouflaged as a coral snake.

Countershading is a form of camouflage in which the top of an animal's body is darker in color, while its underside is lighter. Sharks use countershading. When seen from above, they blend in with the darker ocean water below. This makes it difficult for fishermen—and swimmers—to see them. When seen from below, they blend in with lighter surface water. This helps them hunt because prey species below may not see a shark until it's too late.

Countershading also helps because it changes the way shadows are created. Sunlight illuminates the top of an animal's body, casting its belly in shadow. When an animal is all one color, it will create a uniform shadow that makes the animal's shape easier to see. In countershading, however, the animal is darker where the sun would normally illuminate it, and lighter where it would normally be in shadow. This distorts the shadow and makes it harder for predators to see the animal's true shape.

Creating Camouflage

Animal species are able to camouflage themselves through two primary mechanisms: pigments and physical structures.

Some species have natural, microscopic pigments, known as biochromes, which absorb certain wavelengths of light and reflect others. Species with biochromes actually appear to change colors. Many species of octopus have a variety of biochromes that allow them to change the color, pattern, and opacity of their skin.

Other species have microscopic physical structures that act like prisms, reflecting and scattering light to produce a color that is different from their skin. The polar bear, for instance, has black skin. Its translucent fur reflects the sunlight and snow of its habitat, making the bear appear white.

Camouflage can change with the environment. Many animals, such as the arctic fox, change their camouflage with the seasons. Octopuses camouflage themselves in response to a threat. Other species, such as nudibranchs—brightly colored, soft-bodied ocean "slugs"—can change their skin coloration by changing their diet.

Chameleons change colors in order to communicate. When a chameleon is threatened, it does not change color to blend in to its surroundings. It changes color to warn other chameleons that there is danger nearby.

Some forms of camouflage are not based on coloration. Some species attach or attract natural materials to their bodies in order to hide from prey and predators. Many varieties of desert spiders, for instance, live in burrows in the sandy ground. They attach sand to the upper part of their bodies in order to blend in with their habitat.

Other animals demonstrate olfactory camouflage, hiding from prey by "covering up" their smell or masking themselves in another species' smell. The California ground squirrel, for instance, chews up and spits out rattlesnake skin, then applies the paste to its tail. The ground squirrel smells somewhat like its main predator. The rattlesnake, which senses by smell and body heat, is confused and hesitant about attacking another venomous snake.

Camouflage in Animals

Leaf-tailed Gecko.

Eastern Screech Owl.

Vietnamese Mossy Frog.

Butterfly.

Leaf Katydid.

Mutualism in Animals

In animals, a common mutualistic symbiosis occurs between many herbivores and microorganisms of their digestive tracts. Ungulates (hoofed animals) and some other animals eat plant material that is high in cellulose, even though they lack enzymes capable of breaking down cellulose molecules. They obtain energy from cellulose with the help of symbiotic bacteria and protozoa living within their digestive tracts. These microbes produce enzymes called cellulase that break down cellulose into smaller molecules that the host animal can then utilize. Similarly, wood-consuming termites depend upon symbiotic protozoans living within their intestines to digest cellulose. These are obligate symbioses. The termites cannot survive without their intestinal inhabitants, and the microorganisms cannot live without the host. In each of these symbioses, the host animal benefits from the food provided by the microorganism and the microorganism benefits from the suitable environment and nourishment provided by the host.

A variety of animals engage in a mutualistic relationship referred to as cleaning symbioses. Birds such as oxpeckers benefit their large ungulate hosts by removing their external parasites, benefiting in return from the food source the host provides. In the marine environment, certain species of fish and shrimp similarly specialize in cleaning parasites from the outside of fishes. This mutualistic relationship promotes the well-being of the host fishes and provides food for those that do the cleaning. Unlike herbivores and their gut microorganisms, these interactions do not involve a close association of one organism living exclusively within another. These and other mutualistic but not clearly symbiotic relationships, such as those between plants and their pollinators, are sometimes referred to as proto-cooperation.

Parasitism in Animals

Parasitism is a type of symbiotic relationship, or long-term relationship between two species, where one member, the parasite, gains benefits that come at the expense of the host member.

Examples of Parasitic Animal Relationships

Ticks

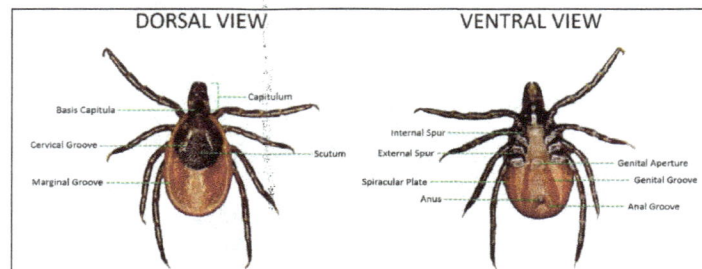

Ticks are arthropod parasites that live on the skin of their animal hosts. Ticks survive by consuming the blood of their hosts, which includes a large variety of animals like dogs, rodents, humans, cattle, and even some lizards. Ticks are attracted to motion, heat, and carbon dioxide, as these are all signs of a suitable host. While not fatal themselves, ticks have the potential to carry and spread more than 10 different pathogens.

Fleas

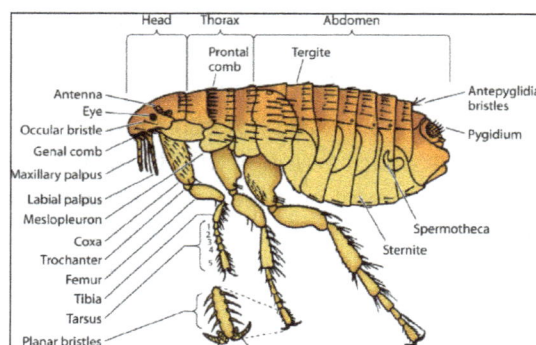

Another common parasitic animal relationship is between the flea and an array of warm-blooded creatures. Depending on the type of flea, this parasite lives off the blood of humans, dogs, cats, rats, and birds. Fleas do not typically transfer disease to their hosts, except for oriental rat fleas which are the primary carrying agent for the plague.

Leeches

Giant Amazon leech (*Haementeria ghilianii*)

crop

salivary glands

proboscis

host's skin

salivary glands

blood vessels

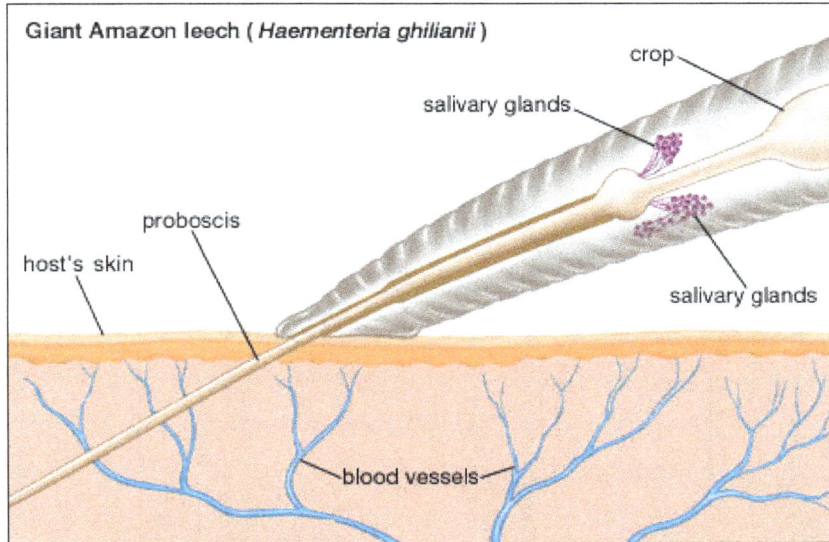

Leeches form parasitic relationships both in and out of water. There are more than 700 different types of leeches, but what most of them have in common is that they live off the blood of almost any animal. In most cases leeches only feed until they are full and then drop from their host, making them a less.

Lice

labrum
clypeus
clypeofrontal suture
frons
postfrontal suture
ocular lobe
prothoracic pleural ridge
mesothoracic pleural ridge
tergal pit
metathoracic pleural ridge
median tergites
spiracle

paratergites

antenna
tibia
eye
femur
trochanter
coxa
thoracic sternum

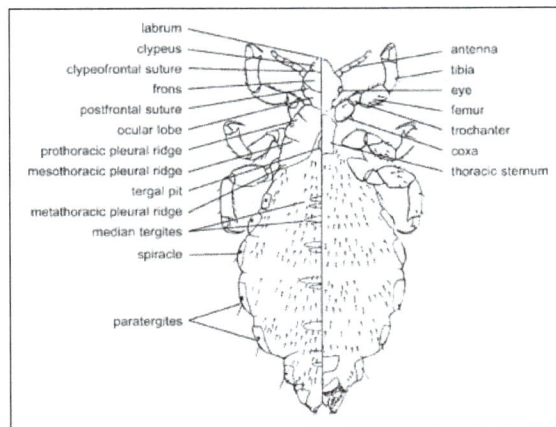

Lice are part of another parasitic animal relationship. In fact, lice find homes on almost every bird and mammal species, including humans. Lice generally act more as scavengers on their host, consuming dead skin cells. However, several types of lice feed on blood or sebaceous fluid, which can cause discomfort to the host.

Helminths

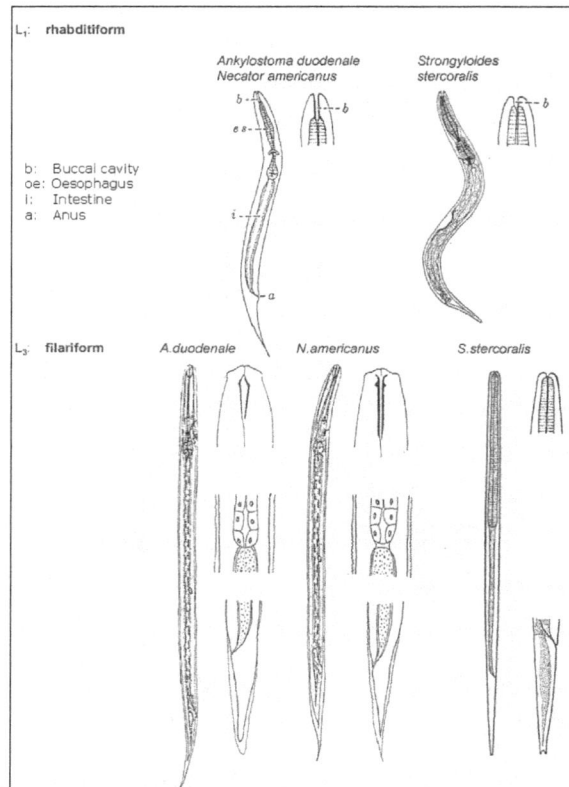

Helminths include parasites like roundworms, flatworms, and thorny-headed worms. Typically helminths live off the digestive tract in mammals, including humans. Helminths cause damage to their host by disrupting the absorption of nutrients in the body, causing weakness, malnutrition and an increased risk for disease and infection.

Predator–Prey Relationships

Predator-prey relations refer to the interactions between two species where one species is the hunted food source for the other. The organism that feeds is called the predator and the organism that is fed upon is the prey.

There are literally hundreds of examples of predator-prey relations. A few of them are the lion-zebra, bear-salmon, and fox-rabbit. A plant can also be prey. Bears, for example, feed on berries, a rabbit feeds on lettuce, and a grasshopper feeds on leaves.

Predators and prey exist among even the simplest life forms on Earth, single-celled organisms called bacteria. The bacteria Bdellovibrio feed on other bacteria that are bioluminescent (they produce internal light due to a chemical reaction). Indeed, the study of Bdellovibrio predation has revealed a great deal of the mechanics of predation and how the predator and prey populations fluctuate in number over time in a related fashion.

Predator and prey populations respond dynamically to one another. When the numbers of a prey such as rabbits explode, the abundance at this level of the food chain supports higher numbers of predator populations such as foxes. If the rabbit population is over-exploited or drops due to disease or some other calamity, the predator population will soon decline. Over time, the two populations cycle up and down in number.

In many higher organisms, the prey can be killed by the predator prior to feeding. For example, a cheetah will stalk, run down, and kill its prey (examples include the gazelle, wildebeest, springbok, impala, and zebra). In contrast, fish and seals that are the prey of some species of shark are examples of prey that is fed on while still alive.

The key aspect of a predator-prey relationship is the direct effect that the predation has on numbers of their prey.

A prey can be vividly colored or have a pattern that is similar to another species that is poisonous or otherwise undesirable to the predator. This sort of strategy, which is known as aposematism, is meant to repel a potential predator based on the predator's previous undesirable experience with the genuine noxious species.

A successful predator must judge when pursuit of a prey is worth continuing and when to abandon the chase. This is because the pursuit requires energy. A predator that continually pursues prey without a successful kill will soon become exhausted and will be in danger of starvation. Predatory species such as lions are typically inactive during the hot daytime hours, when prey is often also resting, but become active and hunt at night when conditions are less energy taxing and prey is more available. Similarly, bats emerge at night to engage in their sonar-assisted location of insects that have also emerged into the air.

When supplied with food in a setting such as a zoo, predators will adopt a sedentary lifestyle. Predation is an energy-consuming activity that is typically done only when the creature is hungry or to supply food for offspring. In settings such as an aquarium, predators and prey will even co-exist.

Being a prey does not imply that the creature is completely helpless. The prey may escape from the predator by strategies such as mimicry, or can simply outrun or hide from the predator. Some species act coordinately to repel a predator. For example, a flock of birds may collectively turn on a predator such as a larger bird or an animal such as a cat or dog to drive off the predator.

This mobbing type of repulsion can be highly orchestrated. For example, when attacked by an animal such as a dog, mockingbirds have been observed to coordinate their attack, with some birds flying close to the animal's face with others pestering it from the rear when it lunges in response. As well, some bird species use different calls, which are thought to be a specific signal to other birds in the vicinity to join the attack. Even birds of a different species may respond to such a call.

The fluctuation in the numbers of a predator species and its prey that occurs over time represents a phenomenon that is known as population dynamics. The dynamics can be modeled mathematically. The results show that a sharp increase in the numbers of a prey species (an example could be a rabbit) is followed soon thereafter by a smaller increase in numbers of the relevant predator (in this case the example could be the fox). As the prey population decreases due to predator killing, the food available for the predators is less, and so their numbers subsequently decline. With the predator pressure reduced, the numbers of the prey can increase once again and the cycle goes on. The result is a cyclical rising and falling of the numbers of the prey population, with a slightly later cyclical pattern of the predator.

Impacts and Issues

Predator-prey relations are an important driving force to improve the fitness of both predator and prey. In terms of evolution, the predator-prey relationship continues to be beneficial in forcing both species to adapt to ensure that they feed without becoming a meal for another predator. This selection pressure has encouraged the development and retention of characteristics that make the individual species more environmentally hardy, and thus collectively strengthens the community of creatures that is part of various ecosystems.

For example, lions that are the fastest will be most successful in catching their prey. Over time, as they survive and reproduce, the number of fast lions in the population will increase. Similarly, the superior attributes that enable prey species to survive will be passed on to succeeding generations. Over time, the fitness of the prey population will also increase. Left to operate naturally, the predator-prey relation will be advantageous for the fitness of both species in relation to how they compete against other species in the same ecosystem. However, since each species improves, their relationship with each other remains unchanged, and the challenge remains to kill or escape from being killed.

The fossil record of Hederellids, which date back almost 400 million years, indicate that the survival race between predator and prey has been a driver of evolution perhaps since evolution began. If so, the predator-prey relationship is fundamentally important to life on Earth.

Predator-prey relationships are also vital in maintaining and even increasing the biological diversity of the particular ecosystem, and in helping to keep the ecosystem stable. This is because a single species is kept under control by the species that uses it for food.

The predator-prey balance of an ecosystem can be disrupted by other changes to the ecosystem including climate related changes such as drought, or human activities that include urban development, foresting, and overuse of resources.

References

- Animal-ecology, social-sciences: encyclopedia.com, Retrieved 4 April, 2019

- Camouflage: nationalgeographic.org, Retrieved 14 February, 2019

- Camouflage-animals: mymodernmet.com, Retrieved 17 June, 2019

- Symbiosis: biologyreference.com, Retrieved 9 August, 2019

- Common-parasitic-animal-relationships: sunnysports.com, Retrieved 29 May, 2019

- Predator-prey-relationships, energy-government-and-defense-magazines, environment: encyclopedia.com, Retrieved 15 March, 2019

Chapter 7

Biodiversity and Conservation

The variety and variability of life on the planet Earth is known as biodiversity. It measures variations on several levels such as genetic, species and ecosystem level. The protection of animals, natural areas and plants is called conservation. This chapter discusses in detail the theories and methodologies related to conservation and biodiversity.

Biodiversity

Biodiversity is the variety of life on earth and includes variation at all levels of biological organisation from genes to species to ecosystems. Genetic, organismal and ecological diversity are all elements of biodiversity with each including a number of components.

Measuring Biodiversity

Purvis & Hector describe three facets of biodiversity that can be measured:

1. Numbers: E.g. the number of genes, populations, species or taxa in an area.

2. Evenness: a site containing 1000 species may not seem very diverse if 99.9% of the species are the same. Many diversity indices have been developed such as Simpson's and Shannon's diversity indices that attempt to convey the extent to which individuals are distributed among species. There are also equivalent measures for genetic diversity such as measures of heterozygosity that incorporate both allele number and relative frequencies.

3. Difference: Some pairs of alleles, populations, species or taxa may be very similar whilst others are very different. For example, if populations within a species are very different they may be considered as different sub-species, management units or evolutionary significant units. Some differences may be considered to be more important than others, for example, ecological differences between species may be important for ecosystem function. All of these kinds of differences are likely to be at least partly reflected by phylogenetic diversity among organisms, which is the sum total of the branch lengths in the evolutionary tree (phylogeny) that links the organisms together. If you sample the phylogeny in different places you will find different things.

Although biodiversity can be measured in lots of different ways the most commonly used measure is that of species richness, there are a number of reasons for this:

1. Species often keep their genes to themselves and thus can have independent evolutionary trajectories and unique histories; it thus makes biological sense to measure species richness rather than a higher taxonomic grouping.

2. It is often easier to count the number of species compared to other measures of biodiversity. Humans tend to be able to recognise species and these are the units typically used in folk knowledge, practical management and political discourse. Humans can visualise variation in biodiversity as variation in species richness.

3. There is a substantial body of information already available on species, for example, in museums and herbaria.

4. Species richness can act as a 'surrogate' for other measures of biodiversity. In general as long as the number of species involved is moderate, greater numbers of species will tend to have more genetic diversity and will tend to have greater ecological diversity as more niches, habitats or biomes will be represented.

There are however some disadvantages in the use of species richness as a measure of biodiversity. One of these is that the number of species that you count depends on the species concept that you adopt. For example, using the biological species concept 40 – 42 species of birds-of-paradise are recognised in Australasia; if the phylogenetic species concept is used this increases to 90 species. Another limitation arises if species richness is used synonymously with biodiversity without emphasising the fact that species richness represents just one element of what biodiversity is.

We also need to consider the spatial scale over which species richness can be considered, with a distinction commonly being made between alpha, beta and gamma diversity. Alpha diversity refers to diversity within a particular area, community or ecosystem and is typically measured as the number of species within that area. Beta diversity is the species diversity between areas and involves comparing the number of species that are unique to each area. Gamma diversity is a measure of the overall diversity across a region.

The diversity of life on Earth has increased over time but it is incredibly difficult to know how many extant species there currently are. Due to the enormity of the task indirect measures are used to estimate the number of extant species and the number varies depending on the assumptions used in the estimation. We are still discovering new species all of the time. A new species of large mammal is still discovered roughly every three years and an average day sees the formal description of around 300 new species across the whole range of life. 13.5 million is a frequently quoted working estimate for the number of species on Earth with a lower estimate of 3.5 and higher estimate of 111.5 million species. Figure shows one estimate of named and unnamed species for a number of taxonomic groups. This highlights the high proportion of species that are as yet unnamed, it also illustrates that diversity is not equally represented within the taxonomic groups; different groups vary widely in their abundance as well as species number.

Mapping Biodiversity

In general as the size of an area increases so does the number of species found within it. This species-area relationship is commonly represented as:

Log S = log c + z log A

S is the number of species,

A is the area

z and c are constants known as the Arrenhius relationship.

Relationships of this type typically explain more than 50% of the variation in species richness between different areas, with the slope of the relationship, z, ranging from 0 to 0.5 (most commonly 0.25 to 0.30). This means that a 90% reduction in the habitat in an area will result in the loss of approximately 50% of the species that live in that habitat, whilst a loss of 99% of the habitat will lead to the extinction of 75% of the species. The constant, z, varies widely however depending on factors such as islands versus continents, latitude and with the range of sizes of areas.

Although larger areas typically contain more species, different areas of the world vary widely in the diversity found within them. A number of conservation organizations such as Conservation International state that these areas should be conservation priorities. There are currently 34 biodiversity hotspots recognized each holding at least 1,500 endemic plant species and having lost 70% of their original habitat area. The hotspots include: Madagascar and the Indian Ocean islands; the coastal forests of Eastern Africa; the Caribbean Islands and the Mediterranean Basin.

Value of Biodiversity

Humans cannot exist without biodiversity as we use it directly and indirectly in a number of ways. Direct use includes things like food, fibres, medicines and biological control, whilst indirect uses includes ecosystem services such as atmospheric regulation, nutrient cycling and pollination. There are also non-use values of biodiversity, such as option value (for future use or non-use), bequest value (in passing on a resource to future generations), existence value (value to people irrespective of use or non-use) and intrinsic value (inherent worth, independent of that placed upon it by humans).

Many of these uses of biodiversity are not incorporated in economic accounts and this leads humans to under-value biodiversity. Ecosystem services and resources such as mineral deposits, soil nutrients, and fossil fuels are capital assets but traditional national accounts do not include measures of the depletion of these resources. This means a country could cut its forests and deplete its fisheries, and this would show only as a positive gain in GDP (gross national product) without registering the corresponding decline in assets (wealth).

The relationship between biodiversity and ecosystem function is clear but a major question in ecology is how much biodiversity is required to maintain ecosystem function. Gaston & Spicer summarize three main ways that ecosystem function can respond to reductions in species richness.

1. Redundancy: There is a minimum number of species required to carry out ecosystem processes and beyond this species are equivalent and their loss of little significance.

2. Rivet-popping: The loss of a few species may have no apparent effect on ecosystem processes but beyond certain thresholds ecosystem services will fail.

3. Idiosyncrasy: Species have complex and varied roles so changes in diversity will cause changes in ecosystem functioning where the direction and magnitude of change is unpredictable.

The design and interpretation of some of the experiments used to test these possible relationships is very contentious but a number of experiments have found relationships that suggest that there is some degree of ecological equivalence between species which fits the redundancy model.

Biodiversity Loss

Biodiversity loss, also called loss of biodiversity is the decrease in biodiversity within a species, an ecosystem, a given geographic area, or Earth as a whole. Biodiversity, or biological diversity, is a term that refers to the number of genes, species, individual organisms within a given species, and biological communities within a defined geographic area, ranging from the smallest ecosystem to the global biosphere. Likewise, biodiversity loss describes the decline in the number, genetic variability, and variety of species, and the biological communities in a given area. This loss in the variety of life can lead to a breakdown in the functioning of the ecosystem where decline has happened.

Bleached coral seascape: A sea turtle swimming over a bleached coral seascape near Heron Island.

The idea of biodiversity is most often associated with species richness (the count of species in an area), and thus biodiversity loss is often viewed as species loss from an ecosystem or even the entire biosphere. However, associating biodiversity loss with species loss alone overlooks other subtle phenomena that threaten long-term ecosystem health. Sudden population declines may upset social structures in some species, which may keep surviving males and females from finding mates, which may then produce further population declines. Declines in genetic diversity that accompany rapid falls in population may increase inbreeding (mating between closely related individuals), which could produce a further decline in genetic diversity.

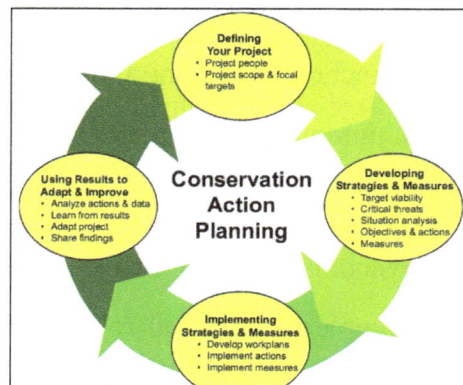

Biodiversity loss: The primary drivers of biodiversity loss are influenced by the exponential growth of the human population, increased consumption as people strive for more affluent lifestyles, and reduced resource efficiency.

Even though a species is not eliminated from the ecosystem or from the biosphere, its niche (the role the species play in the ecosystems it inhabits) diminishes as its numbers fall. If the niches filled by a single species or a group of species are critical to the proper functioning of the ecosystem, a sudden decline in numbers may produce significant changes in the ecosystem's structure. For example, clearing trees from a forest eliminates the shading, temperature and moisture regulation, animal habitat, and nutrient transport services they provide to the ecosystem.

Natural Biodiversity Loss

An area's biodiversity increases and decreases with natural cycles. Seasonal changes, such as the onset of spring, create opportunities for feeding and breeding, increasing biodiversity as the populations of many species rise. In contrast, the onset of winter temporarily decreases an area's biodiversity, as warm-adapted insects die and migrating animals leave. In addition, the seasonal rise and fall of plant and invertebrate populations (such as insects and plankton) which serve as food for other forms of life, also determine an area's biodiversity.

Biodiversity loss is typically associated with more permanent ecological changes in ecosystems, landscapes, and the global biosphere. Natural ecological disturbances, such as wildfire, floods, and volcanic eruptions, change ecosystems drastically by eliminating local populations of some species and transforming whole biological communities. Such disturbances are temporary, however, because natural disturbances are common and ecosystems have adapted to their challenges.

Human-driven Biodiversity Loss

In contrast, biodiversity losses from disturbances caused by humans tend to be more severe and longer-lasting. Humans (Homo sapiens), their crops, and their food animals take up an increasing share of Earth's land area. Half of the world's habitable land (some 51 million square km [19.7 million square miles]) has been converted to agriculture, and some 77 percent of agricultural land (some 40 million square km [15.4 million square miles]) is used for grazing by cattle, sheep, goats, and other livestock. This massive conversion of forests, wetlands, grasslands, and other terrestrial ecosystems has produced a 60 percent decline (on average) in the number of vertebrates worldwide, with the greatest losses in vertebrate populations occurring in freshwater habitats (83 percent) and in South and Central America (89 percent).

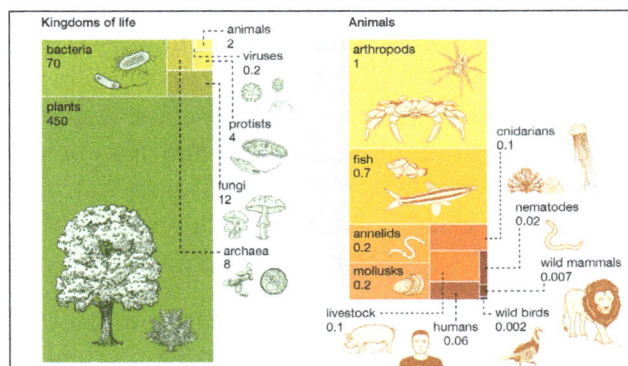

Biomass: Relative biomass on Earth. The planet's biomass is classified by kingdom of life and other major groupings, and the size of each group's relative footprint is displayed using gigatons of carbon as the common measure.

Forest clearing, wetland filling, stream channeling and rerouting, and road and building construction are often part of a systematic effort that produces a substantial change in the ecological trajectory of a landscape or a region. As human populations grow, the terrestrial and aquatic ecosystems they use may be transformed by the efforts of human beings to find and produce food, adapt the landscape to human settlement, and create opportunities for trading with other communities for the purposes of building wealth. Biodiversity losses typically accompany these processes.

There are five important drivers if biodiversity loss:

- Habitat loss and degradation—which is any thinning, fragmentation, or destruction of an existing natural habitat—reduces or eliminates the food resources and living space for most species. Species that cannot migrate are often wiped out.

- Invasive species—which are non-native species that significantly modify or disrupt the ecosystems they colonize—may outcompete native species for food and habitat, which triggers population declines in native species. Invasive species may arrive in new areas through natural migration or through human introduction.

- Overexploitation—which is the harvesting of game animals, fish, or other organisms beyond the capacity for surviving populations to replace their losses—results in some species being depleted to very low numbers and others being driven to extinction.

- Pollution—which is the addition of any substance or any form of energy to the environment at a rate faster than it can be dispersed, diluted, decomposed, recycled, or stored in some harmless form—contributes to biodiversity loss by creating health problems in exposed organisms. In some cases, exposure may occur in doses high enough to kill outright or create reproductive problems that threaten the species's survival.

- Climate change associated with global warming—which is the modification of Earth's climate caused by the burning of fossil fuels—is caused by industry and other human activities. Fossil fuel combustion produces greenhouse gases that enhance the atmospheric absorption of infrared radiation (heat energy) and trap the heat, influencing temperature and precipitation patterns.

Ecologists emphasize that habitat loss (typically from the conversion of forests, wetlands, grasslands, and other natural areas to urban and agricultural uses) and invasive species are the primary drivers of biodiversity loss, but they acknowledge that climate change could become a primary driver. In an ecosystem, species tolerance limits and nutrient cycling processes are adapted to existing temperature and precipitation patterns. Some species may not able to cope with environmental changes from global warming. These changes may also provide new opportunities for invasive species, which could further add to the stresses on species struggling to adapt to changing environmental conditions. All five drivers are strongly influenced by the continued growth of the human population and its consumption of natural resources.

Interactions between two or more of these drivers increase the pace of biodiversity loss. Fragmented ecosystems are generally not as resilient as contiguous ones, and areas clear-cut for farms, roads, and residences provide avenues for invasions by non-native species, which contribute to further declines in native species. Habitat loss combined with hunting pressure is hastening the decline of several well-known species, such as the Bornean orangutan (Pongo pygmaeus), which could become extinct by the middle of the 21st century. Hunters killed 2,000–3,000 Bornean orangutans every year between 1971 and 2011, and the clearing of large areas of tropical forest in Indonesia and Malaysia for oil palm (Elaeis guineensis) cultivation became an additional obstacle to the species' survival. Palm oil production increased 900 percent in Indonesia and Malaysia between 1980 and 2010, and, with large areas of Borneo's tropical forests cut, the Bornean orangutan and hundreds to thousands of other species have been deprived of habitat.

Ecological Effects

The weight of biodiversity loss is most pronounced on species whose populations are decreasing. The loss of genes and individuals threatens the long-term survival of a species, as mates become scarce and risks from inbreeding rise when closely related survivors mate. The wholesale loss of populations also increases the risk that a particular species will become extinct.

Biodiversity is critical for maintaining ecosystem health. Declining biodiversity lowers an ecosystem's productivity (the amount of food energy that is converted into the biomass) and lowers the quality of the ecosystem's services (which often include maintaining the soil, purifying water that runs through it, and supplying food and shade, etc.).

Biodiversity loss also threatens the structure and proper functioning of the ecosystem. Although all ecosystems are able to adapt to the stresses associated with reductions in biodiversity to some degree, biodiversity loss reduces an ecosystem's complexity, as roles once played by multiple interacting species or multiple interacting individuals are played by fewer or none. As parts are lost, the ecosystem loses its ability to recover from a disturbance. Beyond a critical point of species removal or diminishment, the ecosystem can become destabilized and collapse. That is, it ceases to be what it was (e.g., a tropical forest, a temperate swamp, an Arctic meadow, etc.) and undergoes a rapid restructuring, becoming something else (e.g., cropland, a residential subdivision or other urban ecosystem, barren wasteland, etc.).

Reduced biodiversity also creates a kind of "ecosystem homogenization" across regions as well as throughout the biosphere. Specialist species (i.e., those adapted to narrow habitats, limited food resources, or other specific environmental conditions) are often the most vulnerable to dramatic population declines and extinction when conditions change. On the other hand, generalist species (those adapted to a wide variety of habitats, food resources, and environmental conditions) and species favoured by human beings (i.e., livestock, pets, crops, and ornamental plants) become the major players in ecosystems vacated by specialist species. As specialist species and unique species (as well as their interactions with other species) are lost across a broad area, each of the ecosystems in the area loses some amount of complexity and distinctiveness, as the structure of their food chains and nutrient-cycling processes become increasingly similar.

Economic and Societal Effects

Biodiversity loss affects economic systems and human society. Humans rely on various plants, animals, and other organisms for food, building materials, and medicines, and their availability as commodities is important to many cultures. The loss of biodiversity among these critical natural resources threatens global food security and the development of new pharmaceuticals to deal with future diseases. Simplified, homogenized ecosystems can also represent an aesthetic loss.

Economic scarcities among common food crops may be more noticeable than biodiversity losses of ecosystems and landscapes far from global markets. For example, Cavendish bananas are the most common variety imported to nontropical countries, but scientists note that the variety's lack of genetic diversity makes it vulnerable to Tropical Race (TR) 4, a fusarium wilt fungus which blocks the flow of water and nutrients and kills the banana plant. Experts fear that TR4 may drive the Cavendish banana to extinction during future disease outbreaks. Some 75 percent of food crops have become extinct since 1900, largely because of an overreliance on a handful of high-producing crop varieties. This lack of biodiversity among crops threatens food security, because varieties may be vulnerable to disease and pests, invasive species, and climate change. Similar trends occur in livestock production, where high-producing breeds of cattle and poultry are favoured over lower-producing, wilder breeds.

Mainstream and traditional medicines can be derived from the chemicals in rare plants and animals, and thus lost species represent lost opportunities to treat and cure. For example, several species of fungi found on the hairs of three-toed sloths (Bradypus variegatus) produce medicines effective against the parasites that cause malaria (Plasmodium falciparum) and Chagas disease (Trypanosoma cruzi) as well as against human breast cancer.

Solutions to Biodiversity Loss

Dealing with biodiversity loss is tied directly to the conservation challenges posed by the underlying drivers. Conservation biologists note that these problems could be solved using a mix of public policy and economic solutions assisted by continued monitoring and education. Governments, nongovernmental organizations, and the scientific community must work together to create incentives to conserve natural habitats and protect the species within them from unnecessary harvesting, while disincentivizing behaviour that contributes to habitat loss and degradation. Sustainable development (economic planning that seeks to foster growth while preserving environmental quality) must be considered when creating new farmland and human living spaces. Laws that prevent poaching and the indiscriminate trade in wildlife must be improved and enforced. Shipping materials at ports must be inspected for stowaway organisms.

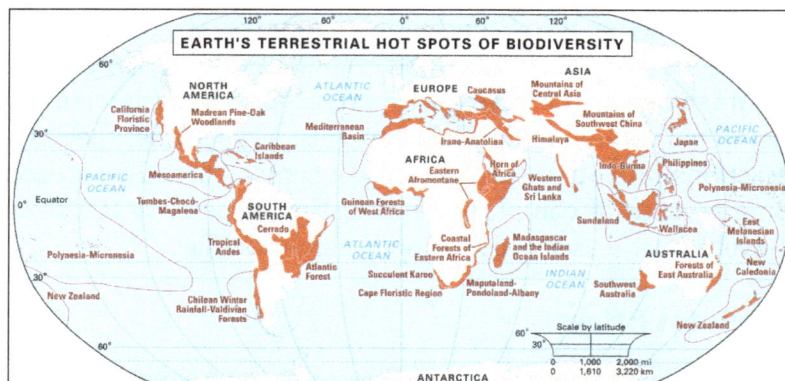

Earth's 25 terrestrial hot spots of biodiversity: As identified by British environmental scientist Norman Myers and colleagues, these 25 regions, though small, contain unusually large numbers of plant and animal species, and they also have been subjected to unusually high levels of habitat destruction by human activity.

Developing and implementing solutions for each of these causes of biodiversity loss will relieve the pressure on species and ecosystems in their own way, but conservation biologists agree that the most effective way to prevent continued biodiversity loss is to protect the remaining species from overhunting and overfishing and to keep their habitats and the ecosystems they rely on intact and secure from species invasions and land use conversion. Efforts that monitor the status of individual species, such as the Red List of Threatened Species from the International Union for Conservation of Nature and Natural Resources (IUCN) and the United States Endangered Species list remain critical tools that help decision makers prioritize conservation efforts. In addition, a number of areas rich in unique species that could serve as priorities for habitat protection have been identified. Such "hot spots" are regions of high endemism, meaning that the species found there are not found anywhere else on Earth. Ecological hot spots tend to occur in tropical environments where species richness and biodiversity are much higher than in ecosystems closer to the poles.

Conservation

Conservation is the study of the loss of Earth's biological diversity and the ways this loss can be prevented. Biological diversity, or biodiversity, is the variety of life either in a particular place or on the entire planet Earth, including its ecosystems, species, populations, and genes. Conservation thus seeks to protect life's variety at all levels of biological organization. Species extinction is the most obvious aspect of the loss of biodiversity.

Factors that cause Extinction

Habitat Loss

Habitat loss is widely listed as the predominant cause of extinction. It is not hard to understand why—clear a forest, destroy a coral reef, or dam a river, and the species found there will likely be lost. These are instances of local extinctions, however, and their occurrence does not mean that the species involved will go extinct everywhere, some scientists use the term extirpation for local extinctions, reserving extinction to mean global extinction.

On Land

In the preceding case histories, the roles of habitat loss in extinction of species are fairly clear-cut. There are seeming counterexamples—Europe and eastern North America, for example—where human actions have extensively modified terrestrial habitats, yet these areas are not centres of extinction. Clearly, habitat destruction causes different numbers of extinctions in different places.

Endemic Species

To learn what makes centres of human-caused extinctions special, one can ask what are their common features. Obviously, many of these places and their species are well known. The 17th-century Dutch artist Rembrandt van Rijn painted birds-of-paradise and marine cone shells gathered from the early European exploration of the Pacific, testifying to people's fascination for attractive and interesting specimens from "exotic" locales. Victorians filled their cabinets with such curiosities—birds, mammals, marine and terrestrial snail shells, and butterflies—and painted tropical flowers and grew them in their hothouses. Yet the natural history of North America and Europe are also very well known. In fact, what is special about the places with so many extinctions is that each area holds a high proportion of species restricted to it. Scientists call such species endemics.

Remote islands have many terrestrial endemics—for instance, more than 90 percent of the plants and land birds of the Hawaiian Islands live nowhere else. Some continental areas are rich in endemic species, too. About 70 percent of the flowering plants in South Africa's fynbos and nearly three quarters of Australia's mammals are endemic. In contrast, many areas have almost no endemic species; only about 1 percent of Europe's birds are found only there.

The simplest model of extinction would be to assume that within a species group—mammals, for instance—all species had roughly the same risk of extinction. Were this to be true, then the more species that live in a region, the more would likely go extinct. This is not the case, however. For example, the Hawaiian Islands and Great Britain in their pasts held very roughly the same number of breeding land birds. Yet, the former have lost more than 100 species, while the latter has lost only a few—and those still survive on the European continent. The difference is that almost all the Hawaiian birds (for instance, honeycreepers such as the apapane and iiwi) were endemic, while only one of Great Britain's birds (the Scottish crossbill) is. Thus, the number of extinctions in an area depends only very weakly on its total number of species but strongly on its total number of endemics. Areas rich in endemic species are where extinctions will concentrate, unless they are so remote that human actions do not harm them.

Apapane (Himatione sanguinea).

Range Size

Given that species differ in their risk of extinction, the size of a species' geographic range is by far the best explanation for the differences. Species with small ranges are much more vulnerable than those with large ranges, simply because it is much easier to destroy the former than the latter.

In regard to range size, the distribution of life on land has several remarkable features. First, many terrestrial species have very small range sizes relative to the average range size. In the Americas, for example, 1 in 10 species of birds and over half the species of amphibians have geographic ranges smaller than the state of Connecticut, and half the bird species have ranges smaller than the states of Washington, Oregon, and California combined. The average range size is very much larger, for some species have huge ranges. The American robin (Turdus migratorius), for example, breeds almost everywhere in the United States from Alaska to Florida to California, across all of continental Canada, and in much of Mexico.

Second, for many kinds of terrestrial organisms, species with small ranges typically have lower local population densities than do widespread species. For example, the American robin is generally a locally common bird across its entire range. But those species with ranges smaller than the size of Connecticut are generally very hard to find even in the midst of their ranges.

Third, terrestrial species with small ranges are geographically concentrated. North America, for instance, has few bird species with small ranges. Such species live almost exclusively in the tropics. The greatest concentrations of bird species are in the Amazon lowlands, with secondary centres in Central America and the forests along the coast of Brazil. But it is the Andes and the coastal forests that have the most bird species with small geographic ranges; i.e., these areas are the centres of endemism. And it is in these areas that threatened species are concentrated.

Terrestrial Hot Spots

In the 1990s a team of researchers led by British environmental scientist Norman Myers identified 25 terrestrial "hot spots" of the world—25 areas on land where species with small geographic ranges coincide with high levels of modern human activity. Originally, these hot spots encompassed about 17 million square km (6.6 million square miles) of the roughly 130 million square km (50 million square miles) constituting Earth's ice-free land surface. Species ranges are so concentrated geographically in these regions that, out of a total of about 300,000 flowering-plant species described worldwide, more than 133,000 occur only there. The comparable numbers for birds are roughly 2,800 of 10,000 species worldwide (of which roughly two-thirds are restricted to the land); for mammals, 1,300 of roughly 5,000 worldwide; for reptiles, roughly 3,000 of about 8,000 worldwide; and for amphibians, 2,600 of roughly 5,000 worldwide.

The hot spots have been sites of unusual levels of habitat destruction. Only about one-eighth of the original habitat of these areas survived to the beginning of the 21st century. Of this remaining habitat, only about two-fifths is protected in any way. Sixteen of the 25 hot spots are forests, most of them tropical forests. For comparison, the relatively less disturbed forests found in the Amazon, the Congo region, and New Guinea have retained about half their original habitats. As a consequence of these high levels of habitat loss, the 25 hot spots are locations where the majority of threatened and recently extinct species are to be found.

Predictions of Extinctions based on Habitat Loss

Worldwide, about 15 percent of the land surface is protected by some form of legislation, though the figure for the 25 hot spots is only 4.5 percent of their original extent. (Such numbers are misleading, however, in that some areas are protected only on paper as their habitats continue to be destroyed.) These statistics lead naturally to the question of how many species will be saved if, say, 4.5 percent of the hot spot land worldwide is protected. Will only 4.5 percent of its species be saved? The answer turns out to be closer to 50 percent, a result that needs some explanation.

The land that countries protect, often as national or regional parks, frequently comprises "islands" of original habitat surrounded by a "sea" of cropland, grazing land, or cities and roads. On a real island the number of species that live there depends on its area, with a larger island housing more species than a smaller one. Many studies involving a wide range of animals and plants show that the relationship between area and species number is remarkably consistent. An island half the size of another will hold about 85 percent of the number of species.

Brazil: The coastal forest of Rio de Janeiro state, Brazil, badly fragmented
as portions were cleared for cattle grazing.

This relationship of species to area holds for "islands" of human-created fragmented habitat. If one were to conduct an experiment to test such an idea, one would take a continuous forest, cut it up into isolated patches, and then wait for species to become locally extinct in them. After sufficient time, one could count the numbers of species remaining and relate them to the area of the patches in which they survived. In the last decades of the 20th century, scientists undertook exactly such an experiment, making use of government-approved forest clearing for cattle ranching, in the tropical forests around Manaus in Brazil. More generally, human actions have repeated this experiment across much of the planet in an informal way. Counts of species in areas of different sizes confirm the species-to-area relationship.

How can these results regarding local habitat "islands" be applied to global extinctions? The mostly deciduous forests of eastern North America provide a case history. The birds of the region have been well-described beginning with the explorations of the naturalist and artist John James Audubon in the first quarter of the 19th century, when the area was still mostly forested. Audubon shot and painted many species including four that are now extinct—the Carolina parakeet, Bachman's warbler (Vermivora bachmanii; see woodwarbler), the passenger pigeon (Ectopistes migratorius), and (if not extinct, then very nearly so) the ivory-billed woodpecker (Campephilus principalis). A fifth species, the red-cockaded woodpecker (Picoides borealis), is endangered. Including the above, about 160 species of birds once lived in these eastern forests.

European settlement cleared these forests, and, at the low point about 1870, only approximately half the forest remained. The species-to-area relationship predicts that about 15 percent of the 160 bird species—that is, about 24 species—would become extinct. Why is this number much larger than the three to five extinctions or near-extinctions observed? The answer can be found by first supposing that all the eastern forests had been cleared, from Maine to Florida and westward to the prairies. Only those species that lived exclusively within these forests—that is, the endemics—would have gone extinct. Species having much larger ranges, such as the American robin mentioned above, would have survived elsewhere. How many species then were originally endemic to the forests of eastern North America? The answer is about 30, the rest having wide distributions across Canada and, for some, into Mexico. The species-to-area predicts that 15 percent of these 30 species, or 4.5 species, should go extinct, which is remarkably close to the observed number.

Passenger Pigeon
(Ectopistes migratorius)

Eastern North America clearly is not a place where species with small ranges are concentrated, but the species-to-area predictions work in other places, too. A variety of studies have examined birds and mammals on the islands of Southeast Asia, from Sumatra westward—one of the biodiversity hot spots (Sundaland). Both here and for birds in Brazil's Atlantic Forest, which is another hot spot, the extent of deforestation and the species-to-area relationship accurately predict the number of threatened species rather than extinct ones. Because the deforestation of these areas is relatively recent, many of their species have not yet become extinct.

The close match between the numbers of extinctions predicted by the species-to-area relationship and the numbers of species already extinct (as in eastern North America) or nearly extinct (as in more recently destroyed habitats) allows simple calculations that have worldwide import. As discussed above, the hot spots retain only 12 percent of their original habitats, and only about 4.5 percent of the original habitats are protected. As predicted by the species-to-area relationship, natural habitats that have shrunk to 4.5 percent of their original extent will lose more than half of their species. Since these habitats once supported 30–50 percent of terrestrial species, very roughly one-fourth of all terrestrial species will likely become extinct.

Species losses will likely be even greater because this calculation does not include nonendemic species—those that live both inside and outside the hot spots. For example, many of the species that live in the relatively less-disturbed tropical moist forests of the Amazon or the Congo region are the same that live in the adjacent hot spots. Human actions are clearing about 10 percent of the original area of these forests every decade, with a half of the area already gone. Species losses in these forests are still relatively few, but the rate will increase rapidly as the last remaining forests dwindle. If only the same percentage of these forests is protected as is the case for the hot spots, then they too will lose half their species.

In summary, many scientists believe that habitat destruction will put somewhere between a fourth and a half of all species on an inexorable path to extinction and will do so within the next few decades. If that proves true, extinction rates by the mid-21st century will be several thousand times the benchmark rate.

Fire Suppression as Habitat Loss

Whereas most of the hot spots are tropical moist forests, four areas—the California Floristic Province, the Cape Floristic Province in South Africa, the Mediterranean Basin, and Southwest Australia—are shrublands. They also are places where people live and grow crops; all four regions are noted for their wines, for example. Not only does this human activity convert land directly, but it also leads to the suppression of fire, especially near people's homes. This alteration of natural fire regimes by the reduction in fire frequencies leads to changes in vegetation, especially to the loss of the native fire-resistant species. Globally, huge areas of grasslands and shrublands would become heavily canopied forests were all fires suppressed. The effects of changes in fire frequencies on species losses have not yet been calculated. Land managers in some fire-adapted habitats have incorporated prescribed fires, which also reduces the available fuel for wildfires, to maintain these natural areas.

In the Oceans

The seas cover more than two-thirds of Earth's surface, yet only 210,000 of the 1,500,000 species that have been described are marine animal, algal, and plant species. Because the oceans are still poorly explored, the count of marine species may be even more of an underestimate than that of land species. For example, the Census of Marine Life, a decadelong international program begun at the start of the 21st century, added 13,000 new marine species to the total count over the first four years of the effort. As on land, the peak of marine biodiversity lies in the tropics. Coral reefs account for almost 100,000 species, yet their total area represents just 0.2 percent of the ocean surface. Between 4,000 and 5,000 species of fish—perhaps a third of the world's marine fish—live on coral reefs. The frequently cited metaphor that "coral reefs are the rainforests of the sea" underscores their importance for conservation.

Coral reef Between 4,000 and 5,000 species of fish—perhaps a third of the world's marine fish—live on coral reefs, some three-fifths of which are threatened by human activities.

When numbers of described marine species are mapped on a worldwide scale, it becomes clear that the global centre of marine biodiversity encompasses the waters of the Philippine and Indonesian islands. Numbers of species drop steeply to the east across the Pacific and less steeply to the west across the Indian Ocean. In the Atlantic Ocean the highest levels of biodiversity are in the Caribbean. Fish, corals, mollusks, and lobsters all show similar patterns in the distributions of their species. Again mirroring the patterns on land, the places with the most species are often not the places with the most endemic species. Major centres of endemism for fish, corals, mollusks, and lobsters include the Philippines, southern Japan, the Gulf of Guinea, the Sunda Islands, and the Mascarene Islands.

With the major exception of the Great Barrier Reef of Australia, most coral reefs are off the coasts of developing countries. Rapidly increasing human populations and poverty put increasing fishing pressure on nearshore reefs. In addition, in their efforts to sustain declining fish catches, people resort to extremely damaging fishing methods such as dynamite and poisons. Coral reefs are also threatened by coastal development, pollution, ocean acidification, and other threats from global warming. Human activities threaten some three-fourths of the world's reefs, with the highest damage being concentrated in areas having high rates of deforestation and high runoff from the land. As the inhabitants of an area destroy their tropical forests, rains erode soils and wash the sediments down rivers into the sea, damaging the local coral reefs. Thus, the destruction of some of the most important terrestrial habitats—in the Caribbean and Southeast Asia—contributes to the destruction of some of the most important marine habitats offshore.

Whereas damage to coral reefs is important for the loss of species, the greatest physical damage to ocean ecosystems involves the effects of bottom trawling, a commercial fishing method making use of a cone-shaped bag of netting that is dragged along the seabed. Damage from bottom trawling occurs over larger areas of Earth than does tropical deforestation, and it involves even greater and more-frequent disturbances, albeit ones not easily seen. Bottom trawling disturbs about 15 million square km (6 million square miles) of the world's seafloor each year. This area of ocean is only about 4 percent of the world total, but its small proportion belies its significance. About 90 percent of the ocean consists of deep waters so poor in nutrients that they are the equivalent of the land's deserts. Almost all of the world's fisheries are concentrated in the 30 million square km (12 million square miles) of nutrient-rich waters that are on the continental shelf, plus a few upwellings. On average, the ocean floor of these productive waters is trawled roughly every two years.

The otter trawl is the most widely used bottom-fishing gear. As it is dragged forward, a pair of flat plates called otter boards—one on each side of the trawl net and weighing several tons—spreads horizontally to keep the mouth of the trawl open; at the same time, a long rope with steel weights keeps the mouth open along its bottom edge. This heavy structure plows the ocean floor as it moves, creating furrows and crushing, burying, and exposing marine life. This activity destroys bottom-dwelling species including corals, brachiopods (lamp shells), mollusks, sponges, sea urchins, and various worms that live on rocks or pebbles on the seabed. It is also damaging to other species, such as polychaete worms that burrow into the seabed. Some species—deep-sea corals, for example—are extremely slow-growing, and they cannot recover before bottom trawls plow the area once again.

Fresh Water

Freshwater ecosystems are divided into two major classes—flowing (such as rivers and streams) and static (such as lakes and ponds). Although the distribution of species in freshwater ecosystems is not as well known as for marine and terrestrial ecosystems, it is still clear that species are similarly concentrated. For fish, the major tropical rivers such as the Amazon River and its tributaries hold a large fraction of the world's freshwater fish species. Tropical lakes, particularly those in the Rift Valley of East Africa, also have large numbers of endemic species.

Riverine habitats have been extensively modified by damming and by channelization, the latter being the practice of straightening rivers by forcing them to flow along predetermined channels. A global survey published in the early 21st century revealed just how few of the world's large rivers are natural. In the contiguous United States almost every large river has been modified extensively. Much the same is true in Europe. In terms of the area of their basins, more than half of the world's rivers are extensively modified. The water of some rivers barely reaches its final destination; this is the case for the Colorado River in the United States, which empties into the Gulf of California, and for the Amu Darya in Central Asia, which empties into the rapidly shrinking Aral Sea. Along their routes the water is used for agriculture or lost as evaporation from dams. Large wild rivers are typical only of Arctic regions in Alaska, Canada, and Siberia—places so far away from urban centres that there has been no incentive to control their waters. The massive changes to the world's rivers explain why such large fractions of species living in rivers have become extinct or may do so soon.

Pollution

Pollution is a special case of habitat destruction; it is chemical destruction rather than the more obvious physical destruction. Pollution occurs in all habitats—land, sea, and fresh water—and in the atmosphere. Global warming, is one consequence of the increasing pollution of the atmosphere by emissions of carbon dioxide and other greenhouse gases.

Water pollution is a global-scale problem, no less so for rivers and marine life. Wastes are often dumped into rivers, and they end up in estuaries and coastal habitats, regions that support the most diverse shallow-water ecosystems and the most productive fisheries. Rivers receive pollution directly from factories that dump a wide variety of wastes into them. They also receive runoff, which is rainwater that has passed over and through the soil while moving toward the rivers. In fact, water entering rivers after it has been used for irrigation has passed through the soil more than once—first as runoff, which is then returned to the land for irrigation, whereupon it soaks through the soil again, often picking up fertilizers and pesticides.

Some polluted river water eventually reaches freshwater wetlands. In the case of the Florida Everglades, runoff from the agricultural areas upstream adds unwanted nutrients to an ecosystem that is naturally nutrient-poor. As it does so, the vegetation changes, and species not common in the Everglades begin to take over the natural habitats. To prevent this, agricultural water is often discharged to the ocean instead, which deprives the Everglades of much needed water and can trigger or augment toxic red tides and algal blooms in coastal waters.

Other polluted waters reach estuaries on the way to the oceans; estuaries are among the most polluted ecosystems on Earth. On entering the oceans, the polluted waters can harm the ecosystems there. The Mississippi River, for example, drains a basin of more than 3 million square km (1.2 million square miles), delivering its water, sediments, and nutrients and other pollutants into the northern Gulf of Mexico. Fresh water is less dense than salt water and floats on top. This upper layer contains the nitrogen and phosphorus fertilizers that have run off croplands, and they fertilize the ocean's phytoplankton, causing excessive population growth. As the masses of phytoplankton die, sink, and decompose, they deplete the water's oxygen. Bottom dwellers such as shrimp, crabs, starfish, and marine worms suffocate and die, creating a "dead zone." In 2017 the Gulf of Mexico's dead zone reached 22,730 square km (8,776 square miles)—an area the size of New Jersey.

Such conditions also occur in Europe's Baltic, Adriatic, and Black seas. The Baltic has gone from being naturally nutrient-poor and diverse in species to being nutrient-rich and degraded in its ecosystems within a few decades. In the Adriatic Sea, rising nutrient levels have generated a large increase in phytoplankton. Nutrients in the runoff flowing into the Black Sea seem to be contributing factors in the invasion and subsequent massive increase since the 1980s of the comb jelly Mnemiopsis leidyi.

This has caused the decline of native species and fisheries.

Similar nutrient enrichment has led to increasing frequencies of toxic blooms of microscopic organisms such as Pfiesteria piscicida in the eastern United States, a dinoflagellate that kills fish and has been reported to cause skin rashes and other maladies in humans.

Rising levels of pollution may have also contributed to a wave of outbreaks of diseases affecting marine life. Caribbean coral reefs have been particularly affected, with successive waves of disease propagating throughout the region in recent decades. The result has been large declines in two species of major reef-building corals, Acropora cervicornis and A. palmata, and the herbivorous sea urchin Diadema antillarum. Their combined loss has transformed Caribbean reefs from high-coral, low-algae ecosystems to high-algae, low-coral ones. The latter type of ecosystems supports far fewer species.

Introduced Species

Often implicate introduced species as a cause of species extinctions. Humans have spread species deliberately as they colonized new areas, just one example being the Polynesians as they settled the eastern Pacific islands. New Yorkers in the 1890s wanted all the birds in Shakespeare's works to inhabit the city's Central Park, and they introduced the starling (Sturnus vulgaris) to North America as a consequence. Through the centuries hunters have demanded exotic birds and mammals to shoot, fishermen have wanted challenging fish, and gardeners have wanted beautiful flowers. Nonetheless, the consequences in some cases have been devastating. Cacti and the shrub Lantana camara, for example, which were introduced as ornamental plants, have destroyed huge areas of grazing land worldwide.

Common starling (Sturnus vulgaris).

Not all introductions are deliberate. Several species of rats, for instance, were likely unwelcome hitchhikers on ocean voyages. Humans have spread infectious diseases around the world with consequences that dramatically changed human history; a recent example is AIDS, spread from Africa by individuals infected with the human immunodeficiency virus (HIV). Introduction of the brown tree snake that destroyed Guam's birds was also accidental.

Almost every type of living organism has been moved deliberately or accidentally. A study published in the 1990s surveyed species that traveled in the ballast tanks of oceangoing ships. Such ships take on seawater as ballast in one port and release it in other ports sometimes half a world away. In one example, water taken aboard in Japan and released in Oregon contained 360 species including microscopic single-celled animals and plants, jellyfish and corals, mollusks, various kinds of marine worms, and fish—indeed, an almost complete catalog of major kinds of marine organisms.

Although not all species devastate the communities they enter, a closer look at what happened on Guam after the introduction of the brown tree snake (Boiga irregularis) illustrates just how damaging a species can be. Birds started disappearing from the central region of the island in the early 1960s. At the same time, the snakes started causing blackouts as they crawled up power poles and short-circuited the suspended lines, and many people who kept chickens noticed that snakes were killing the birds in their coops. Before Boiga made its appearance, 10 species of native birds lived in the forests of Guam—the Micronesian starling (Aplonis opaca), the Mariana crow (Corvus kubaryi), the Micronesian kingfisher (Halcyon cinnamomina), the Guam flycatcher (Myiagra freycineti), the Guam rail (Rallus owstoni), the rufous fantail (Rhipidura rufifrons), the bridled white-eye (Zosterops conspicillatus), the white-throated ground dove (Gallicolumba xanthonura), the Mariana fruit dove (Ptilinopus roseicapilla), and the cardinal honeyeater (Myzomela cardinalis). An 11th species, the island swiftlet (Aerodramus vanikorensis), nested in caves and fed over the forest. In addition, the island supported a few introduced bird species, birds that nested in wetland habitats, and some seabirds. The 10 forest birds declined following similar patterns—first on the central part of the island, where the main town and port are, and then on southern Guam by the late 1960s. By 1983 scientists could find the birds only in a small patch of forest in the north of the island, and by 1986 they were gone from there as well. The nonnative birds suffered a similar pattern of decline, and many of the waterbirds and seabirds that nested on the island also declined. The starling has survived on a small island, Cocos, off the south coast of Guam, and the swiftlet still nests on the high walls of a cave. The snake has eliminated every species for which it could easily kill the eggs and young in the nest. Of the 10 forest species, the rail and the flycatcher were endemic to the island, while the kingfisher is a distinctive subspecies. The kingfisher and the rail survive in captivity.

Brown tree snake (Boiga irregularis).

Islands such as Guam seem to be particularly vulnerable to introduced species. People have often brought domestic cats with them to islands, where some of the animals have escaped to form feral populations. Cats are efficient predators on vertebrates that have had no prior experience of them; for example, cats have caused the extinction of some 30 island bird species worldwide. Introduced herbivores, particularly goats, devastate native plant communities, again because the plants have had no chance to adapt to these new threats.

Rufous fantail (Rhipidura rufifrons).

Although islands are particularly vulnerable, introduced species also wreak large-scale changes on continents. An exceptionally clear example is the loss of the American chestnut (Castanea dentata) in eastern North America after the accidental introduction of the fungus, Endothia parasitica, that causes chestnut blight. The chestnut, a once-abundant tree, was removed with surgical precision over its entire range beginning in the early 20th century. All that survive are small individuals that tend to become infected as they get older.

As briefly mentioned above, hybridization is another mechanism by which introduced species can cause extinction. In general, species are considered to be genetically isolated from one another— they cannot interbreed to produce fertile young. In practice, however, the introduction of a species into an area outside its range sometimes leads to interbreeding of species that would not normally meet. If the resident species is extremely rare, it may be genetically swamped by the more abundant alien. One example of a species threatened by hybridization is the white-headed duck. The European population of this species lives only in Spain, where habitat destruction and hunting once reduced it to just 22 birds. With protection, it recovered to about 800 individuals, but it is now threatened by a related species, the ruddy duck (O. jamaicensis). This bird is native to North America, was introduced to Great Britain in 1949, and spread to the continent including Spain, where it hybridizes with the much rarer white-headed duck. Similar cases involve other species of ducks, frogs, fish, cats, wolves, and other groups of organisms worldwide.

Overharvesting

Overharvesting, or overfishing in the case of fish and marine invertebrates, depletes some species to very low numbers and drives others to extinction. In practical terms, it reduces valuable living resources to such low levels that their exploitation is no longer sustainable. Whereas the most-familiar cases involve whales and fisheries, species of trees and other plants, especially those valued for their wood or for medicines, also can be exterminated in this way.

Whaling

Whaling offers an example of overharvesting that is interesting not only in itself but also for demonstrating how poorly biodiversity has been protected even when it is of economic value. The first whalers likely took their prey close to shore. Right whales were the "right" whales to take because they are large and slow-moving, feed near the surface and often inshore, float to the surface when harpooned, and were of considerable commercial value for their oil and baleen. The southern right whale (Eubalaena australis), for example, is often seen in shallow, sheltered bays in South Africa and elsewhere. Such behaviour would make any large supply of raw materials a most tempting target. Whalers had nearly exterminated the North Atlantic species of the northern right whale (Eubalaena glacialis) and the bowhead whale (Greenland right whale; Balaena mysticetus) by 1800. They succeeded in exterminating the Atlantic population of the gray whale (Eschrichtius robustus). Whalers then moved on to species that were more difficult to kill, such as the humpback whale (Megaptera novaeangliae) and the sperm whale (Physeter macrocephalus).

Humpback whale (Megaptera novaeangliae) breaching.

Japanese factory ship hauling a minke whale through a slipway in the ship's stern.

The Napoleonic Wars gave whales a respite, but with the peace of 1815 came a surge of whalers into the Pacific Ocean, inspired by the stories of James Cook and other explorers. The first whalers arrived in the Hawaiian Islands in 1820, and by 1846 the fleet had grown to nearly 600 ships, the majority from New England. The catch on each whaling voyage averaged 100 whales, though a voyage could last as long as four years.

In the late 1800s, steamships replaced sailing ships, and gun-launched exploding harpoons replaced hand-thrown lances. The new technology allowed whalers to kill what until then had been the "wrong" whales—fast-swimming species such as the blue whale (Balaenoptera musculus) and fin whale (B. physalus). Whalers killed nearly 30,000 blue whales in 1931 alone; World War II gave the whales a break, but the catch of blue whales rose to 10,000 in 1947. The fin whale was next, with the annual catch peaking at 25,000 in the early 1960s; then came the smaller sei whale (B. borealis)—which no one had bothered to kill until the late 1950s—and finally the even smaller minke whale (B. acutorostrata), which whalers still hunt despite an international moratorium in effect since 1986 that seeks to curb commercial whaling.

The story of whaling is, in brief, the rapid depletion and sometimes extermination of one population after another, starting with the easiest species to kill and progressing to the most difficult. That whales are economically valuable raises the obvious question of why there were no attempts to harvest whales sustainably.

Fishing

Overfishing is the greatest threat to the biodiversity of the world's oceans, and contemporary information published for fisheries in the United States can serve as an example of the magnitude of the problem. Congress requires the National Marine Fisheries Service (NMFS) to report regularly on the status of all fisheries whose major stocks are within the country's exclusive economic zone, or EEZ. (Beyond its territorial waters, every coastal country may establish an EEZ extending 370 km [200 nautical miles] from shore. Within the EEZ the coastal state has the right to exploit and regulate fisheries and carry out various other activities to its benefit.) The areas involved are considerable, covering portions of the Atlantic, the Caribbean, the Gulf of Mexico, and the Pacific from off San Diego to the Bering Sea out to the west of the Hawaiian island chain along with the islands constituting the western part of the former Trust Territory of the Pacific Islands. At the turn of the 21st century, the NMFS deemed some 100 fish stocks to be overfished and a few others close to being so, while some 130 stocks were not thought to be overfished. For about another 670 fish stocks, the data were insufficient to allow conclusions. Thus, a little under half of the stocks that could be assessed were considered overfished. For the major fisheries—those in the Atlantic, the Pacific, and the Gulf of Mexico—two-thirds of the stocks were overfished.

As to the hundreds of stocks about which fisheries biologists know too little, most of them are not considered economically important enough to warrant more investigation. One species, the barndoor skate (Raja laevis), was an incidental catch of western North Atlantic fisheries in the second half of the 20th century. As the name suggests, this is a large fish, too big to go unrecorded. Its numbers fell every year, until by the 1990s none were being caught, and it was listed as an endangered species.

Logging and Collecting

Similar cases of overharvested species are found in terrestrial ecosystems. For example, even when forests are not completely cleared, particularly valuable trees such as mahogany may be selectively logged from an area, eliminating both the tree species and all the animals that depend on it. Another example is the coast sandalwood (Santalum ellipticum), a tree endemic to the Hawaiian Islands that was almost completely eliminated from its habitats for its wood and fragrant oil. Rosewood (various species) is used in fine furniture and is the most trafficked wild item.

Some species are overharvested not to be killed but to be kept alive and sold as pets or ornamental plants. Many species of parrots worldwide, for example, are in danger because of the pet trade, and the survival of cacti and orchid species is threatened by collectors.

Secondary Extinctions

Once one species goes extinct, there will likely be other extinctions or even an avalanche of them. Some cases of these secondary extinctions are simple to understand—e.g., for every bird or mammal that goes extinct, one or several species of parasite also will likely disappear. From well-studied species, it is known that bird and mammal species tend to have parasites on or inside them that can live on no other host.

Other extinctions cause changes that can be quite complicated. Species are bound together in ecological communities to form a food web of species interactions. Once a species is lost, those species that fed on it, were fed on by it, or were otherwise benefited or harmed by that species will all be affected, for better or worse. These species, in turn, will affect yet other species. Ecological theory suggests that the patterns of secondary extinction are quite complicated and thus may be difficult to demonstrate.

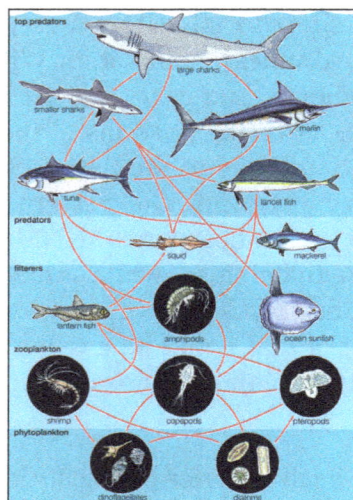

Generalized aquatic food web showing the network of feeding relationships that exist among species in a marine community. The disappearance of one species will affect other species up, down, and across the web, for better or worse.

The most easily recognizable secondary extinctions should be seen in species that depend closely on each other. In the Hawaiian Islands, anecdotal evidence for secondary extinctions comes from a consideration of nectar-feeding birds. Before modern human activity on the islands, there were three nectar-feeding Hawaiian honeycreepers—the mamo (Drepanis pacifica), the black mamo (Drepanis funerea), and the iiwi (Vestiaria coccinea)—that had long decurved (downward-curving) beaks, the kind adapted to inserting into appropriately long and curved flowers. The first two birds are extinct, whereas the third is extinct on two islands, is very rare on a third, and has declined on others.

Mamo (Drepanis pacifica), a nectar-feeding honeycreeper.

Extensive habitat destruction is likely the cause of many species of Hawaiian birds. In addition, native Hawaiians hunted some species for their feathers. In the case of the three honeycreepers described above, however, their extinctions may have followed the destruction of important nectar-producing plants by introduced goats and pigs. Many of the native lobelias, such as those of the genera Trematolobelia and Clermontia, have clearly evolved to be pollinated by the three honeycreepers, and the plants were once important components of the forest's understory. About a quarter of these plant species are now extinct, a rate that plainly exceeds those for the rest of the flora, perhaps because they were so vulnerable to introduced mammalian herbivores. It is not certain, however, whether the plants disappeared first and then their bird pollinators or vice versa.

Similarly, some surviving Hawaiian birds seem to be unusually specialized feeders and to be threatened as a consequence of the loss of their food sources. For example, another rare honeycreeper, the akiapolaau (Hemignathus munroi), is an insectivore that feeds on insects mainly on large koa (Acacia koa) trees. Today, however, few koa forests remain, because the trees have been overharvested for their attractive wood. Yet another Hawaiian honeycreeper, a seed-eating species called the palila (Loxioides bailleui), is endangered because it depends almost exclusively on the seeds of one tree, the mamane (Sophora chrysophylla), which is grazed by introduced goats and sheep.

Stories of secondary extinctions are nearly always unsatisfactory anecdotes because of the difficulty of teasing apart the various explanations for what happened in the past. There is abundant evidence from small-scale ecological experiments that a change in one species' numbers (including its complete elimination) will cause cascading effects in the abundance of other species. The stories are plausible enough that particular attention should be paid to the future consequences of contemporary extinctions.

Global Warming

The global effects of flooding the atmosphere with carbon dioxide and other greenhouse gases created as by-products of human activity are many and complex. Global warming, an increase in global average surface temperature, is but one of them. Both the land and the oceans are increasing in temperature at different rates in different places. The Arctic and Antarctic seem to be heating up the most, while temperature changes in the tropics are more modest. In addition, the heating is melting glaciers and ice sheets, causing sea levels to rise, and it may be increasing the frequency of the most intense hurricanes and the intensity of other storms and extreme weather, with consequences to the flow of rivers. Some areas and their ecosystems are becoming wetter, while others are becoming drier and hotter and thus likely to suffer more wildfires. Scientific knowledge is constantly increasing and changing what is known about the way Earth works, and knowledge of global warming and its effects on species is expanding particularly quickly.

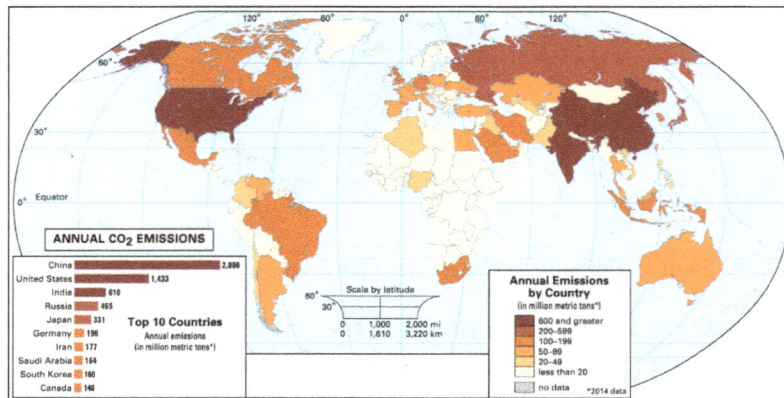

Carbon dioxide emissions Map of annual carbon dioxide
emissions by country in 2014.

Since the turn of the 21st century, the relationship between changes to Earth's physics and chemistry and biodiversity has been clarified significantly. Although the precise effects of global warming on species extinction rates are still uncertain, they almost certainly will be large.

It is now clear that most species are shifting their geographic ranges toward cooler places and are starting important events such as breeding, migration, and flowering earlier in the year. One study, published in the early 21st century, found that more than 80 percent of nearly 1,500 species of animals and plants from a wide variety of habitats worldwide were changing in the direction expected from global warming. Another study, published about the same time, found consistent northward movement of the northern boundaries of animal and plant species in Europe and North America. Such geographic shifts of species alter important ecological interactions with their prey, predators, competitors, and diseases. Some species can benefit, but others can be harmed—for example, when migratory insect-eating birds arrive too late to exploit the emergence of moth larvae that they typically feed to their young.

Even if scientists know that species likely will move in the direction of cooler habitats—toward the poles or up mountainsides—they will find the exact changes difficult to predict because so many different factors are involved. For example, one contemporary study examined the changes in the ranges and feeding ecology of butterflies in Great Britain in the last decades of the 20th century. Two species were found to have increased in range, as expected, but by unexpected amounts. One species lived in isolated habitat patches and could not fly the distance to some of the patches that were suitable for populating. Rising temperatures increased the number and density of suitable patches, allowing the butterfly to reach the distant patches by making use of intervening newly available ones as "stepping-stones." The second butterfly species was able to exploit a previously unused food plant that grew in shady places that formerly had been too cool. With that change in diet, the butterfly was able to greatly expand its range. In both cases, the species benefited from the warming climate, at least at the northern edges of their ranges. It takes little imagination to foresee the consequences if, for example, the mosquito Aedes aegypti, which carries dengue hemorrhagic fever, were to expand its range to an unexpected degree across the southern United States, which is its present northern limit.

Can anything useful be inferred about the likely consequences of climate change to species extinction? Assuming that the change has simple effects, scientists can predict where a species should be in the future if it is to live in the same range of climatic conditions that it does now. The real concern is that this range of suitable conditions, or the species' climate envelope, may shrink to nothing as conditions change—i.e., there may be no suitable conditions for a species in the future.

Other things being equal, one would expect that species with small geographic ranges will more likely be affected than species with large ranges. For a species with a large range, global warming may cause the species to disappear from the south and appear farther north, but the change in the size of the range may be quite small. The species may be present in many of the same areas both before and after global warming. The American robin, could be expected to behave this way.

Of the species that have very small ranges, many live in mountainous areas, as can be seen in the map. To date, human activity has had relatively small effects on such species because of the impracticalities of cultivating land in mountains, especially on steep slopes. But it is these species—living on cool mountaintops that are now becoming too warm for them—that have nowhere else to go. Such arguments lead scientists to believe that extinctions caused by rising temperatures will be additional to those caused by land-use changes such as deforestation. Rough calculations suggest that rising temperatures may threaten about a quarter of the species in hot spots.

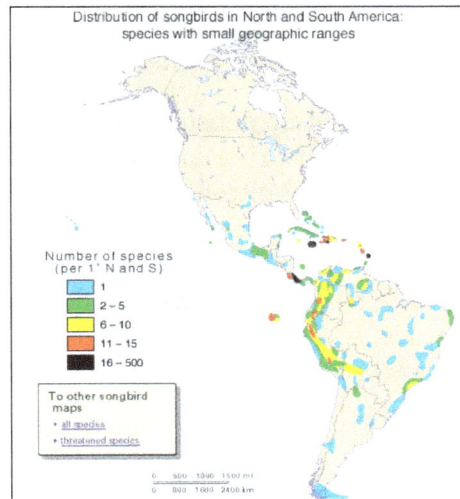

Distribution of songbirds in North and South America: species with small geographic ranges

Number of species (per 1° N and S)

	1
	2 – 5
	6 – 10
	11 – 15
	16 – 500

To other songbird maps
- all species
- threatened species

0 500 1000 1500 mi
0 800 1600 2400 km

Which Species are most Vulnerable to Extinction

Before a species becomes extinct, it must first be rare. Some species are much more vulnerable than others to becoming dangerously rare, and other species, when rare, are more likely than others of equal rarity to succumb.

Baikal seals (Phoca sibirica), endemic to Lake Baikal, southeastern Siberia, Russia.

Endemism and Rarity

A small geographic range makes a species particularly vulnerable to global extinction. Many of the threats to species are geographically restricted, so species with large ranges will survive somewhere even if they are locally extirpated.

The collections of plants and animals on islands are often rich in endemic species; those species that are restricted to particular islands must inevitably have small ranges. Since 1600 the great majority of known bird extinctions have been on islands. The problem of extinction is not just limited to islands.

Body Size and Related Reproductive Characteristics

Generally, the larger the body size of an animal, the longer it lives and the fewer offspring it produces each year. Relatively large animals also tend to have relatively low population densities; thus, a viable population of, say, elephants occupies considerably more space than an equal-sized population of rabbits. Large predators such as tigers (Panthera tigris) have lower population densities than the herbivores on which they feed. A tiger has a home range that may occupy 100 square km (40 square miles), while a rabbit may survive in 1 hectare (0.004 square mile).

Large animals, by virtue of their low population densities, are at increased risk of extinction. Moreover, an animal species that produces few offspring each year and that suffers a major loss in numbers from human activity will need more time to recover than a species with high reproductive rates. Again taking the tiger as an example, a tiger population large enough to avoid all the perils of low population size (including the chance that all offspring will be the same sex or that genetic inbreeding will occur) requires extremely large areas. Although the tiger is not considered an endemic species—it has a very large geographic range, stretching over much of eastern and southern Asia— its rarity in that range means that it is nonetheless vulnerable to extinction from human activity.

Behavior

Size of Home Range

In the case of lions, wolves, and wild dogs, population size alone is a poor predictor of their local extinction, even when the animals live in protected areas. For these species, conflict with people on the borders of the protected areas is the major cause of mortality. Border areas are population "sinks," areas where the death rate of the population exceeds the birth rate and into which individuals enter through migration from more-central portions of the protected area. This explains, for example, why African wild dogs are under particular threat—more so than other species, such as cheetahs and leopards, that also are afforded protected areas. Wild dogs live in large packs that roam very large areas, in contrast to the two cat species, which are mostly solitary and which have smaller home ranges. The more widely a species moves (i.e., the larger its home range), the more likely it is to move beyond areas where it is protected.

Concentration

Some species have aggregation behaviours that make them vulnerable to disturbance or hunting. For example, bats may congregate in large numbers in particular caves to have their young, making significant portions of their total population especially susceptible when their habitat is disturbed by human visitation or damaged by the cave's commercialization or flooding. The various species of groupers often come together to spawn on a few nights of each year tied to phases of the moon and at traditional mating sites. Fishermen who know these sites and the timing of spawning can devastate large populations of these species by concentrating their efforts during this most vulnerable time in the fish's life cycle.

Low Dispersal

Because small populations are so much more at risk than large ones, individuals of species that can readily disperse can rescue local populations on the verge of extinction. The previously discussed example of the checker spot butterfly illustrates, first, the rescue of some small populations by the dispersal of individuals from larger "reservoir" populations nearby and, second, the subsequent extinction of the reservoirs as urban development isolated them from still other butterfly populations that could have rescued them. Species that have low dispersal rates are at a disadvantage because it is unlikely that one population can save another.

Mating Systems

Small populations suffer from inbreeding, an inevitable tendency of mating individuals in a small isolated population to be more closely related than they would be in a larger one. When population size is severely reduced, inbreeding may be the final insult that will cause the remaining population to go extinct. The likelihood that this will happen, however, seems related to the social structure of the species involved.

The northern elephant seal (Mirounga angustirostris) of the Pacific Coast of North America was thought to have been hunted to extinction in the late 1800s, though it later became apparent that perhaps 20–30 individuals persisted locally for a couple of decades before the population began to recover gradually under protection. The Indian rhinoceros (Rhinoceros unicornis) in the early 20th century was reduced to two isolated populations—one numbering between 12 and 100, the other between 60 and 80—before protection allowed it to make a limited recovery. Moreover, not all of the rhinoceros males in the reduced population were likely to have bred. Today the elephant seal is genetically uniform suggesting that a high degree of inbreeding occurred during the time its population was at a minimum, whereas the rhino has probably lost little of its genetic variability. The population histories of the two species are similar.

Northern elephant seal (Mirounga angustirostris).

The social structures of elephant seals and rhinos are dramatically different. Each year, the one dominant bull seal that guards the harem is likely to father all of the young. An isolated seal population thus may become genetically uniform relatively quickly because very few males father each generation of young. Rhinos, on the other hand, are largely monogamous, so a group of them will have a greater number of fathers than a comparably sized group of seals.

Insularity

The vulnerability of island species is likely a combination of two factors—their endemism and rarity and their ecological naivete, the latter being exemplified by the greater effect of domestic cat introductions on unwary island bird species than on more "streetwise" mainland species. Nevertheless, some island bird species are less likely to be threatened than similar bird species found on continents. The reason lies in the abundance of island species—they are often quite numerous on their islands, for they have fewer competitors than do mainland species.

Human use

Many species are hunted for meat and other products, including whales and various fish, as discussed above. Less familiar is the widespread trade in bushmeat, which is essentially everything that can be hunted—from mice to chimpanzees and gorillas—and is especially prevalent in West and Central Africa. Yet other species are harvested for body parts, such as tiger bones and rhino horns, which are used in Asian medicines. A wide variety of plants are harvested too, again often for medicinal purposes. Simply put, any species that is used for food, wood, or medicine or as pets or houseplants, that is collected (such as butterflies or invertebrate shells), or that attracts attention for any other reason suffers an increased risk of extinction.

Preventing the Loss of Biodiversity

A thorough knowledge of the factors that cause extinction and the vulnerability of different species to them is an essential part of conserving species. In large part, conservation is about removing or reducing those factors and doing so for the most vulnerable species and in the places where species are most vulnerable. Much of the task of conservation professionals is to protect habitats large enough to house viable populations of species, first deciding where the priorities should be and sometimes restoring habitats that already have been destroyed. Local conservation groups often spend time removing introduced species, which can mean physically weeding invasive plants or trapping invasive animals. These activities must be accompanied by efforts to prevent introductions of new threats. Others work to reduce harvesting directly or to reduce the incidental catch of nontarget species. Nonetheless, there are a variety of specific tools that can be applied to different circumstances, as categorized below and illustrated by case histories. Sometimes, when working with very rare species, scientists may not know the exact causes of threat, which can lead to intense arguments about exactly how to proceed.

Species Interventions

Protective Custody

Some species become so rare that there are doubts about whether they will be able to survive in the wild. Under such circumstances, the species may be brought into protective custody until areas can be made suitable for their release back into the wild.

Protective custody is an important tool in plant conservation, where a large number of seeds can be easily stored. In addition, botanic gardens can grow rare plants, protecting the species until such times as they can be planted in the wild. Of a total of about 300,000 described species of flowering plants, Botanic Gardens Conservation International has estimated that about 80,000 species are protected in botanic gardens and a few thousand additional species in other facilities—together about one-third of the total. Kew Garden's Millennium Seed Bank—with the mission to conserve 25 percent of the world's bankable plant species by 2020—became the largest wild plant seed bank in the world. By 2018 it contained about 13 percent of the world's wild plant species, holding some 2.25 billion seeds from 189 countries.

For animals, zoos provide an important refuge for some vertebrate species but not for the vast majority of animals, which are invertebrates. Of species protected in zoos, a number of them have later been returned to the wild. To do so, however, substantial problems need to be overcome.

Determining that a species should be brought into captivity and then deciding what to do with the individuals are illustrated by the California condor. The first key decision was whether to bring the birds into protective custody or to manage their small population in the wild. There was no dispute that the condors were once widespread, ranging across the southern and western United States and northern Mexico. As long ago as the turn of the 20th century, however, they were restricted to the mountains of southern California. The widespread practice of setting out poisoned carcasses to kill livestock predators was likely the major cause of their decline. Exactly how many condors still survived was at the core of the debate. Some thought their numbers had been declining constantly, from 150 in the 1950s to 60 in 1970. Others posited constant numbers, and, according to one opinion, if there was no decline, there was need for neither explanation nor intervention—the condors, though rare, should be left alone.

Eventually, photographic surveys completed the catalog of individuals, removing doubts about which individuals were alive, which were dead, and why. Studies of nesting showed that those birds that bred did so with reasonable success. Radio-collared birds showed that the species foraged far beyond the remote areas of its nesting sites, so simply leaving the birds alone and protecting the habitat near the nest would not be sufficient. Birds died from ingesting lead in animals that had been shot or the toxic substances in poison-laced carrion. In the 1980s all the remaining birds were brought into captivity, although not without a lawsuit over the issue.

Because captive populations are almost always small, there is a high risk of inbreeding. Thus, the condors were carefully screened genetically to ensure that as much as possible of the genetic variability of the species was preserved. The captive-breeding program was ultimately a success and between 12 and 20 fledglings were produced each year after 1991. Earlier production had been lower, likely as a consequence of the birds' inexperience in nesting and rearing. Efforts to teach captive-reared birds through artificial means—some nestlings were fed by puppets resembling condor heads—were much less successful. These birds were particularly inept when released, as real parents teach their young many important things about living in the wild. Eventually, scientists released birds in Arizona, California, and Baja California, and some of these birds have reared young of their own in the wild.

Genetic Intervention

In small populations, inbreeding can cause genetic variability to be lost quite quickly. A simple example is provided by the Y chromosome in humans (and other mammals), which confers maleness and which behaves like human surnames do in large parts of the world. If every human couple had just two children each generation, then by chance alone 25 percent of the couples would have two sons, each with one Y chromosome from the father; 25 percent would have two daughters, each with no Y chromosome; and the remaining 50 percent would have one son and one daughter. The surnames of the fathers with two daughters would be lost as they married and had their own children, as indeed would be those fathers' Y chromosomes. In a small breeding population, after just a few generations, every individual would have the last name of the same male ancestor—perhaps a great-great-grandfather—and the same Y chromosome.

Many genetic mutations are deleterious, reducing the individual's chances of survival, but are also recessive, requiring the inheritance of one mutant gene from each parent to manifest their effect. This means that in a large population, their effect will be masked by the overwhelming numerical superiority of the normal dominant gene. For the reasons explained above, in a small population chance events cause the loss of genetic variability and so increase the likelihood that individuals will suffer harmful genetic diseases.

A study on mammals kept in zoos illustrates the harmful effects of inbreeding. In the past, to maintain sufficient productivity, zookeepers often bred animals that proved to be good at producing young. Because of this practice, some breeding pairs quickly came to have the same grandparents and, in some cases, the same parents. The studies showed that such pairs produced young that were much less likely to survive than young from pairs of unrelated individuals. Many modern zoos serve as "gene banks" and store genetic reserves of their animals. Zoos in various locations around the world can exchange the semen of endangered species for artificial insemination to promote genetic diversity. Minimizing inbreeding is especially critical for the preservation of animals that are extinct in the wild or critically endangered.

The practical problem for conservation is whether to place efforts on genetic intervention—bringing in "new blood," that is, individuals and so genes, from the outside—or to concentrate on the factors causing the initial decline.

This issue was at the heart of the management dilemma posed by the Florida panther (Puma concolor coryi), a distinct subspecies of puma (P. concolor) confined to a small, isolated, and inbred population in southern Florida. The specific question was whether to introduce pumas from Texas into the Florida population. Florida panthers once had been part of a continuous widespread population. In the 19th century they became isolated in southern Florida. As their numbers declined, the occurrence of genetic defects increased, including sperm and heart defects and undescended testicles in males. Conservation scientists hoped that the introduction of pumas from outside Florida would reverse the genetic damage. This proposal was highly controversial; as with the example of the California condor, some argued that the population was doing well in its limited range and did not need additional animals. Despite concerns, in the mid-1990s eight females from Texas were released, and scientists closely followed the fates of them, of their young, and of the young of cats with only Florida parents. It was found that, although hybrid cats—those with both Florida and Texas parents—do not seem to live longer than pure Florida cats and hybrid females do not produce more kittens than pure Florida females, hybrid kittens survive about twice as well. Since the introduction of the Texas females, numbers of Florida panthers have increased, and hybrid cats were expanding the known range of their habitats.

Florida panther (Puma concolor coryi).

Protecting Species

For species that are hunted or collected, direct protection may be an essential conservation tool. National laws, such as the Endangered Species Act in the United States, make collecting or killing an endangered species or threatened species illegal. An example of such a protected species in the United States is the country's national bird, the bald eagle (Haliaeetus leucocephalus). International laws protect whales of various species, and such agreements as the Convention on International Trade in Endangered Species prohibit commercial trade in designated species.

Bald eagle (Haliaeetus leucocephalus).

The problems of implementing protection are illustrated by the conservation of the two African species of rhinoceros. The population of the black rhinoceros (Diceros bicornis) fell to about 2,400 individuals in 1995, down from a likely number of several hundred thousand at the start of the 20th century, when it ranged over most of southern Africa. The white rhinoceros (Ceratotherium simum) historically had a smaller geographic range. Today its northern subspecies occurs only in the Democratic Republic of the Congo, where it is very rare. The southern subspecies lives almost entirely in South Africa, Namibia, Zimbabwe, and Kenya. Together, they numbered under 12,000 in 2001, again likely a small fraction of their original numbers.

White rhinoceros (Ceratotherium simum), Mkuze Game Reserve, South Africa.

Although conversion of habitat land to agricultural use and sport hunting caused the earlier rhino declines, the major threat is now poaching—entering reserves where the animals are protected and killing them for their horns, which are in high demand in parts of the world for dagger handles and, in powdered form, as an ingredient of traditional medicines. The costs of protecting rhinos are considerable. WWF has estimated that it costs $1,400 per square km ($3,600 per square mile) per year to detect and deter poaching. Reserves that hold rhinos often cover tens of thousands of square kilometres; the multimillion-dollar budgets required are beyond the means of many African governments.

Removing Invasive Species

Many conservation programs have tackled invasive species, and, once again, the message from the two case studies that follow is that, although these programs can be successful, they are often expensive.

Introduced domestic cats have caused the extinction of many island species. In the 1970s scientists estimated that the cat population on Marion Island, one of the two Prince Edward Islands in the subantarctic Indian Ocean, was killing 450,000 seabirds each year, jeopardizing the birds' survival. The cats, which were estimated to number at least 2,000, were descended from five animals brought to the island in 1949 to deal with a mouse problem at a meteorological station there. Islands are small and isolated, and because cats are predators, they tend to have much smaller populations than their prey. Nonetheless, it took 19 years of an intensive eradication program to remove the last cats.

The second case study is the purple loosestrife (Lythrum salicaria), a plant that has overrun thousands of square kilometres of North American wetlands, replacing the naturally diverse vegetation of grasses, sedges, and other wetland plants. It is native to Europe and was introduced into North America in the early 1800's. It now occurs across most of the continental United States and is most prevalent in the northeastern and north-central part of the country and in Canada.

Like other exotic weeds, purple loosestrife can spread rapidly. Control measures have included the use of herbicides and the release of herbivorous insects, including different species of beetles that eat the leaves, roots, and seeds of the plant. The latter strategy, an instance of biological control, is not risk-free. It involves introducing a second nonnative species to control a nonnative species introduced earlier. The effects of the insect introductions have not always been clear, and, as with other examples of biological control, there is a danger that the introduced herbivorous insects will harm native plant species.

Introducing Species

The introductions of Texas pumas into the Florida panther population and of captive-reared condors back into parts of their original habitat were successful, as discussed above. So were the introductions to North America of the starling, and the house sparrow (Passer domesticus), which was introduced to New York City from Europe in the 1850s. What is often overlooked, however, is that many other attempts have failed. This in fact is the typical result.

For example, after extensive efforts in the 19th and early 20th centuries to introduce nonnative game birds to the United States, it was found that, even with releases of large numbers of birds, most of the attempts failed. An exception has been the common, or ring-necked, pheasant (Phasianus colchicus), native to China and introduced to the United States in the 1890s. This low rate of success has an important implication—even when it is known from hindsight that an individual introduction can succeed, as did the pheasant, most such introductions still will fail. When returning species to the wild that have already gone extinct there, the prognosis is even bleaker. One reason in some cases is that the cause of the extinctions—for instance, the brown tree snake responsible for bird extinctions on Guam is still present.

Habitat Protection

Because the loss of habitat is the primary reason that species are lost both locally and globally, protecting more habitat emerges as the most important priority for conservation. This simple idea raises difficult questions. Which habitats should be protected? And because it seems unlikely that all habitats can be protected, which ones should receive priority.

If reserves were judiciously placed over the identified hot spots of biodiversity, the special places where vulnerable species are concentrated, a large fraction of species might be saved. Presently, the allocation of reserves around the world is poor. Reserves larger than 100,000 square km (40,000 square miles) are generally in high mountains, tundras, and the driest deserts, areas that are not particularly species-rich. On the other hand, hot spots such as Madagascar and the Philippines protect less than 2 percent of their land.

The same kinds of questions hold on smaller scales, as illustrated by a study reported in the late 1990s. The Agulhas Plain at the southern tip of Africa is one of the world's "hottest" spots for concentrations of vulnerable plant species. An area only some 1,500 square km (600 square miles) in size was found to house 1,751 plant species, 99 of them endemic. Whereas most of the state forests and private nature reserves in the area are coastal, most of the hot spot's endemic plants live inland. Given that new reserves must be created if these plants are to survive, where should they be situated to encompass the maximum number of species at minimum cost.

Fortunately, the data available to make these decisions included a knowledge of the distribution of plant species over the Agulhas Plain in fairly good detail—the kind of information not likely to be available in most hot spots. This allowed the plain's plant-species composition to be divided into a grid of cells, each 3 × 3 km on a side. Computer algorithms (systematic problem-solving methods) were then used to select sets of cells from the grid according to their complementary species composition—that is, the aim was to encompass as many species or as many endemics as possible in as small an area (as few grid cells) as possible.

Naively applied, these algorithms will not give useful results. For example, the sites they select may not be available for reserves. Also, the choice of too small a cell size can lead to the selection of protected areas containing populations so small and widely scattered that they would be unlikely to persist. This is fittingly dubbed the "Noah's Ark effect," because the ark held only two individuals of each species for a short time. Reserves need to be large enough to support species indefinitely. The choice of a cell size of 3 × 3 km is politically feasible because reserves of this size already have been established in the region and are probably ecologically sensible for many plant species. Other factors had to be included in the final selection of cells. Some areas are unsuitable for various reasons—for instance some are overrun by invasive plants, while others are mostly in urban areas or croplands. In contrast, other areas are particularly desirable—for example, they may be adjacent to existing reserves, and it is easier to expand such reserves than to create new ones. The results of this study thus provided advice for establishing reserves that combined ecological information on species distribution with practical and political considerations.

Saving the most species for the least money likewise was the consideration that motivated another study published in the late 1990s, of which counties in the United States should be conservation priorities. An earlier study that attempted to locate sites for new reserves in the United States had equated efficiency with the minimum number of counties needed to achieve a given coverage of endangered species. That approach would have been sensible if land were much the same price everywhere. Unfortunately, the study's targets had included counties encompassing San Diego, Santa Cruz, and San Francisco in California, Honolulu in Hawaii, and certain counties in Florida, all of which are among the highest-priced land in the country. The later study asked how many species could be protected for a given total cost. It found that considerable savings in cost per species accrue from selecting larger, more-complementary areas and lower total costs and that, as a consequence of this approach the places identified for protection were often quite different from those recommended in the earlier study.

Habitat Connections

Habitats that are not completely destroyed may be fragmented to the degree that individual fragments are too small to hold viable populations of many species—they may suffer from inbreeding or the increased demographic risks. Yet in total the fragments may actually be of sufficient area to support these species. An obvious conservation intervention is to find ways of connecting fragments by wildlife corridors. These corridors can be created from currently unprotected land between existing reserves or by restoring land between existing habitat fragments.

Many contemporary efforts to create corridors are small-scale; they can be as simple as hedgerows that connect woodlots, a strategy that likely works for some small species. Other efforts are far more ambitious. One of the earliest large efforts involved a plan to connect various parks and other protected areas in Florida by corridors of land that would have to be purchased or otherwise protected from development. From this initial effort, a conservation group, the Wildlands Project, has developed an extensive set of plans for many areas in North America intended to set priorities for the acquisition of land for a mosaic of corridors that would eventually link together large parts of the continent.

A regional example of corridor creation is in Costa Rica. Situated in the Caribbean lowlands, La Selva Biological Station is one of the major centres for research on tropical forests. Occupying an area of about 16 square km (6 square miles), the station is bordered on the south by Braulio Carrillo National Park, a much larger area of forest covering 460 square km (180 square miles). The national park extends to La Selva through a forest corridor that descends in elevation from nearly 3,000 metres (nearly 10,000 feet) at Barva Volcano down to 35 metres (115 feet) above sea level at La Selva. The corridor had been threatened by agricultural development until conservation groups, realizing that La Selva would become an isolated forest "island," purchased the corridor to protect it.

Habitat Management

Once protected, areas must often be managed in order to maintain the threatened species within them. Management may involve the removal of alien species, as previously discussed. It can also involve restoring natural ecological processes to the area. Original fire and flooding regimes are examples of such processes, and they are often controversial because human actions can alter them significantly.

Fire Control

Despite the often valid reasons for suppressing wildfires, the practice can change vegetation dramatically and sometimes harm species in the process. Human activities have changed fire regimes across large areas of the planet, including some biodiversity hot spots. Getting the fire regimes right can be essential for conserving species.

An example of a species for which the control of fire regimes has proven both possible and essential is Kirtland's warbler. This endangered species nests only in the Upper Peninsula of Michigan, an exceptional case of a bird species with a tiny geographic range well outside the tropics. The bird places its nest in grasses and shrubs below living branches of jack pines (Pinus banksiana) that are between 5 and 20 years old. The region's natural wildfires originally maintained a sufficient area of young jack pines. As elsewhere, modern practices suppressed fires, and the habitat declined. The birds are also susceptible to cowbirds, which are parasitic egg-layers. Active management with prescribed fires to ensure that there are always jack pines of the right age, together with the removal of cowbirds, has steadily increased the population of warblers since the early 1990s, when it had included only about 200 singing males. By 2015 there were about 2,360 singing males.

Flood Control

In much the same way that human actions suppress fire regimes, they also control water levels, and the resulting changes can have important consequences for endangered species. An example of a species so affected is the Cape Sable seaside sparrow (Ammodramus maritimus mirabilis) found in the Florida Everglades. The Everglades once stretched from Lake Okeechobee in the north to Florida Bay in the south. Water flowed slowly over a wide area, and its levels varied seasonally: summer rains caused the levels to rise in late summer and early fall; then the dry season dropped the levels to their lowest in late May. Under this natural regime, some areas were continuously flooded for many years, drying out in only the driest years, whereas others were flooded for only a few months each year. It is in these drier prairies that the sparrows nest from about the middle of March until the water floods their nests in the summer.

Water-management actions have diverted the flow of water to the west of its natural path, making the western part of the Everglades unnaturally wet in some years during the bird's nesting season. Unnatural flooding during the four years that followed 1992 reduced the bird population in the west to less than 10 percent of its 1992 level. In the east the prairies have become unnaturally dry and so have become subject to an increased number of fires, which has jeopardized the sparrow population there. It is clear that rescuing the Cape Sable seaside sparrow requires the restoration of the natural water levels and flows within the Everglades and the consequent return to natural fire regimes.

Habitat Restoration

Once a habitat has been destroyed, the only remaining conservation tool is to restore it. The problems involved may be formidable, and they must include actions for dealing with what caused the destruction. Restorations are massive ecological experiments; as such, they are likely to meet with different degrees of success in different places. Restoration of the Everglades, for example, requires restoring the natural patterns of water flow to thousands of square kilometres of southern Florida.

A case history of habitat restoration comes from the Midwestern United States. In Illinois, natural ecosystems cover less than 0.1 percent of the state, so restoration is almost the only conservation tool available. North Branch is a 20-km (12-mile) strip of land running northward from Chicago along the north branch of the Chicago River. Early in the 20th century, it was protected from building but later abandoned. Beginning in the 1970s, a group of volunteers first cleared out introduced European buckthorn (Rhamnus cathartica), removed abandoned cars, and planted seeds that they had gathered from combing the tiny surviving remnants of original prairies in places such as along railways and in cemeteries. The initial effort to replant the prairie species failed, however; animals ate the small growing plants. The solution was to restore the natural fire regime, although controlled burns—setting fires safely near homes—posed a difficult technical challenge. But once it was accomplished, the results were immediate and dramatic. The original prairie plants flourished and the weeds retreated, although with an important exception. Under the native trees grew non-native thistles and dandelions.

The original habitats locally called barrens constituted a visually striking and ecologically special habitat. Restoring them was a particular challenge, and the main conservation problem was finding the right mix of species. One recommendation was to use remnant barrens as models, but the North Branch volunteers rejected them as being too degraded. In the early 20th century, naturalists had speculated that barrens were special because they lacked some characteristic prairie species and, at the same time, had their own distinctive species. The volunteers examined habitat descriptions in old treatises of local plants looking for such species, many of which certainly would now be scarce. The key discovery was a list of barrens plants published by a country doctor in 1846. The scientific names had changed in 150 years, but, by tracking them through the local literature, the volunteers found many of their putative barrens-specific plants on the doctor's list. In possession of this vital information, the group succeeded in restoring barrens sites that by 1991 held 136 native plants, including whole patches consisting of species that, until restoration, had been locally rare.

The example above illustrates two rules of ecological restoration. The first is that one must be able to save all the component species of the original ecological community. The second which is illustrated by the doctor's list, is that one needs to know which species belong and which do not. Without the right species mix, restorers must constantly weed and reseed.

Permissions

We would like to thank the editorial team for lending their expertise to make the book truly unique. They have played a crucial role in the development of this book. Without their invaluable contributions this book wouldn't have been possible. They have made vital efforts to compile up to date information on the varied aspects of this subject to make this book a valuable addition to the collection of many professionals and students.

This book was conceptualized with the vision of imparting up-to-date and integrated information in this field. To ensure the same, a matchless editorial board was set up. Every individual on the board went through rigorous rounds of assessment to prove their worth. After which they invested a large part of their time researching and compiling the most relevant data for our readers.

The editorial board has been involved in producing this book since its inception. They have spent rigorous hours researching and exploring the diverse topics which have resulted in the successful publishing of this book. They have passed on their knowledge of decades through this book. To expedite this challenging task, the publisher supported the team at every step. A small team of assistant editors was also appointed to further simplify the editing procedure and attain best results for the readers.

Apart from the editorial board, the designing team has also invested a significant amount of their time in understanding the subject and creating the most relevant covers. They scrutinized every image to scout for the most suitable representation of the subject and create an appropriate cover for the book.

The publishing team has been an ardent support to the editorial, designing and production team. Their endless efforts to recruit the best for this project, has resulted in the accomplishment of this book. They are a veteran in the field of academics and their pool of knowledge is as vast as their experience in printing. Their expertise and guidance has proved useful at every step. Their uncompromising quality standards have made this book an exceptional effort. Their encouragement from time to time has been an inspiration for everyone.

The publisher and the editorial board hope that this book will prove to be a valuable piece of knowledge for students, practitioners and scholars across the globe.

Index

www.ingramcontent.com/pod-product-compliance
Lightning Source LLC
Chambersburg PA
CBHW082021190326
41458CB00010B/3232